TEXTBOOK OF ANALYTICAL CHEMISTRY

TEXTBOOK OF ANALYTICAL CHEMISTRY

By

Dr. Syed Aftab Iqbal
M.Sc. , Ph.D., FICS
FICC, FIAEM, MNASc.
Professor
Department of Chemistry
Saifia Science College
Barkatullah University
Bhopal (India)

&

Dr. Ameera Al Haddad
Assistant Professor
Department of Chemistry
University of Bahrain
BAHRAIN

DISCOVERY PUBLISHING HOUSE PVT. LTD.
NEW DELHI-110 002

Published by:
Namit Wasan
DISCOVERY PUBLISHING HOUSE PVT. LTD.
4383/4B, Ansari Road, Darya Ganj
New Delhi-110 002 (India)
Phone : +91-11-23279245; 23253475; 43596065
E-mail : discoverybooksindia@gmail.com
discoverypublishinghouse@gmail.com
namitwasan9@gmail.com
web : www.discoverypublishinggroup.com

Edition: 2020

ISBN: 978-81-8356-830-2

Textbook of Analytical Chemistry

Printed at:
Infinity Imaging Systems
Delhi

Preface

Analytical chemistry is the study of the chemical composition of natural and artificial materials. Properties studied in analytical chemistry include geometric features such as molecular morphologies and distributions of species, as well as features such as composition and species identity. Unlike the sub disciplines inorganic chemistry and organic chemistry, analytical chemistry (like physical chemistry) is not restricted to any particular type of chemical compound or reaction. The contributions made by analytical chemists have played critical roles in the sciences ranging from the development of concepts and theories (pure science) to a variety of practical applications, such as biomedical applications, environmental monitoring, quality control of industrial manufacturing and forensic science (applied science).

Analytical chemistry is a sub discipline of chemistry that has the broad mission of understanding the chemical composition of all matter and developing the tools and experiments to make either qualitative or quantitative measurements. In a nutshell, analytical chemistry applies measurement science principles along with an understanding of chemical systems to provide useful information; and it has significant overlap with other branches of chemistry through the measurement methods that it provides. For example, the field of bioanalytical chemistry is a growing area of analytical chemistry that addresses analytical questions in biochemistry, (the chemistry of life). Experimental physical chemistry and analytical chemistry show similarity in that both have a measurement science focus. While there is occasional overlap between these two disciplines, the goal of physical chemistry experiments is to determine the dynamics of how energy affects the chemical system by making kinetic, thermodynamic, and spectroscopic measurements, while analytical chemistry sometimes uses such energy-based measurements to determine matter-related properties such as chemical identity and quantity. Analytical chemistry is also focused on improvements in experimental design and the creation of new measurement tools to provide better chemical information.

Traditionally, analytical chemistry was particularly concerned with the questions of "what chemicals are present, what are their characteristics and in what quantities are they present?" These questions are often involved in questions that are more dynamic such as what chemical reaction an enzyme catalyzes or how fast it does it, or even more dynamic such as what is the transition state of the reaction. The next logical steps of understanding what it means, how it fits into a larger system, how can this result be generalized into theory, or how it can be used all result from information provided by analytical chemistry methods.

Authors

Contents

CHAPTER - 1

Introduction

Analytical chemistry is the study of the chemical composition of natural and artificial materials. Properties studied in analytical chemistry include geometric features such as molecular morphologies and distributions of species, as well as features such as composition and species identity. Unlike the sub disciplines inorganic chemistry and organic chemistry, analytical chemistry (like physical chemistry) is not restricted to any particular type of chemical compound or reaction.The contributions made by analytical chemists have played critical roles in the sciences ranging from the development of concepts and theories (pure science) to a variety of practical applications, such as biomedical applications, environmental monitoring, quality control of industrial manufacturing and forensic science (applied science).

Analytical chemistry is a sub discipline of chemistry that has the broad mission of understanding the chemical composition of all matter and developing the tools and experiments to make either qualitative or quantitative measurements. In a nutshell, analytical chemistry applies measurement science principles along with an understanding of chemical systems to provide useful information; and it has significant overlap with other branches of chemistry through the measurement methods that it provides. For example, the field of bioanalytical chemistry is a growing area of analytical chemistry that addresses analytical questions in biochemistry, (the chemistry of life). Experimental physical chemistry and analytical chemistry show similarity in that both have a measurement science focus. While there is occasional overlap between these two disciplines, the goal of physical chemistry experiments is to determine the dynamics of how energy affects the chemical system by making kinetic, thermodynamic, and spectroscopic measurements, while analytical chemistry sometimes uses such energy-based measurements to determine matter-related properties such as chemical identity and quantity. Analytical chemistry is also focused on improvements in experimental design and the creation of new measurement tools to provide better chemical information.

Traditionally, analytical chemistry was particularly concerned with the questions of "what chemicals are present, what are their characteristics and in what quantities are they present?" These questions are often involved in questions that are more dynamic such as what chemical reaction an enzyme catalyzes or how fast it does it, or even more dynamic such as what is the transition state of the reaction. The next logical steps of understanding what it means, how it fits into a larger system, how can this result be generalized into theory, or how it can be used all result from information provided by analytical chemistry methods.

Analytical chemists work to improve the reliability of existing techniques to meet the demands for better chemical measurements which arise constantly in our society. They adapt proven methodologies to new kinds of materials or to answer new questions about their composition. They carry out research to discover completely new principles of measurement and are at the forefront of the utilization of major discoveries such as lasers and microchip devices for practical purposes. They make important contributions to many other fields as diverse as forensic chemistry, archaeology, and space science.

MODERN ANALYTICAL CHEMISTRY

Modern analytical chemistry is dominated by instrumental analysis. Many analytical chemists focus on a single type of instrument. Academics tend to either focus on new applications and discoveries or on new methods of analysis. The discovery of a chemical present in blood that increases the risk of cancer would be a discovery that an analytical chemist might be involved in. An effort to develop a new method might involve the use of a tunable laser to increase the specificity and sensitivity of a spectrometric method. Many methods, once developed, are kept purposely static so that data can be compared over long periods of time. This is particularly true in industrial quality assurance (QA), forensic and environmental applications. Analytical chemistry plays an increasingly important role in the pharmaceutical industry where, aside from QA, it is used in discovery of new drug candidates and in clinical applications where understanding the interactions between the drug and the patient are critical.

HISTORY

Much of early chemistry was analytical chemistry since the questions of what elements and chemicals were present in the world around us and what are their fundamental natures is very much in the realm of analytical chemistry. There was also significant early progress in synthesis and theory which of course are not analytical chemistry. During this period significant analytical contributions to chemistry include the development of systematic elemental analysis by Justus von Liebig and systematized organic analysis based on the specific reactions of functional groups. The first instrumental analysis was flame emissive spectrometry developed by Robert Bunsen and Gustav Kirchhoff who discovered rubidium (Rb) and caesium (Cs) in 1860.

Most of the major developments in analytical chemistry take place after 1900. During this period instrumental analysis becomes progressively dominant in the field. In particular many of the basic spectroscopic and spectrometric techniques were discovered in the early 20th century and refined in the late 20th century. The separation sciences follow a similar time line of development and also become increasingly transformed into high performance instruments In the 1970s many of these techniques began to be used together to achieve a complete characterization of samples. Starting in approximately the 1970s into the present day analytical chemistry has progressively become more inclusive of biological questions (bioanalytical chemistry), whereas it had previously been largely focused on inorganic or small organic molecules. Lasers have been increasingly used in chemistry as probes and even to start and influence a wide variety of reactions. The late 20th century also saw an expansion of the application of analytical chemistry from somewhat academic chemical questions to forensic, environmental, industrial and medical questions, such as in histology.

Traditionally, analytical chemistry has been split into two main types, qualitative and quantitative:

QUALITATIVE

- Qualitative inorganic analysis seeks to establish the presence of a given element or inorganic compound in a sample.
- Qualitative organic analysis seeks to establish the presence of a given functional group or organic compound in a sample.

QUANTITATIVE

Quantitative analysis seeks to establish the amount of a given element or compound in a sample.

APPROACHES

Most modern analytical chemistry is categorized by two different approaches such as analytical targets or analytical methods. Analytical Chemistry (journal) reviews two different approaches alternatively in the issue 12 of each year.

BY ANALYTICAL TARGETS

- Bioanalytical Chemistry
- Material analysis
- Chemical analysis
- Environmental analysis
- Forensics

BY ANALYTICAL METHODS

- Spectroscopy
- Mass Spectrometry
- Spectrophotometry and Colorimetry
- Chromatography and Electrophoresis
- Crystallography
- Microscopy
- Electrochemistry

TRADITIONAL ANALYTICAL TECHNIQUES

Although modern analytical chemistry is dominated by sophisticated instrumentation, the roots of analytical chemistry and some of the principles used in modern instruments are from traditional techniques many of which are still used today. These techniques also tend to form the backbone of most undergraduate analytical chemistry educational labs. Examples include:

TITRATION

Titration involves the addition of a reactant to a solution being analyzed until some equivalence point is reached. Often the amount of material in the solution being analyzed may be determined. Most familiar to those who have taken college chemistry is the acid-base titration involving a color changing indicator. There are many other types of titrations, for example potentiometric titrations. These titrations may use different types of indicators to reach some equivalence point.

GRAVIMETRY

Gravimetric analysis involves determining the amount of material present by weighing the sample before and/or after some transformation. A common example used in undergraduate education is the determination of the amount of water in a hydrate by heating the sample to remove the water such that the difference in weight is due to the water lost.

INORGANIC QUALITATIVE ANALYSIS

Inorganic qualitative analysis generally refers to a systematic scheme to confirm the presence of certain, usually aqueous, ions or elements by performing a series of reactions that eliminate ranges of possibilities and then confirms suspected ions with a confirming test. Sometimes small carbon containing ions are included in such schemes. With modern instrumentation these tests are rarely used but can be useful for educational purposes and in field work or other situations where access to state-of-the-art instruments are not available or expedient.

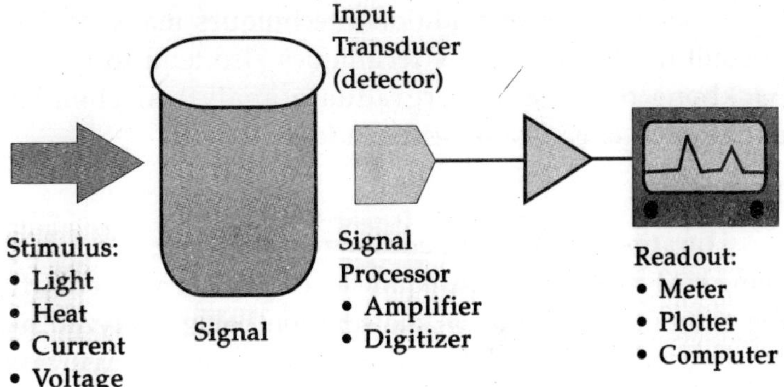

Fig. 1.1: Analytical instrument showing the stimulus and measurement of response

INSTRUMENTAL ANALYSIS

Spectroscopy

Spectroscopy measures the interaction of the molecules with electromagnetic radiation. Spectroscopy consists of many different applications such as atomic absorption spectroscopy, atomic emission spectroscopy, ultraviolet-visible spectroscopy, x-ray fluorescence spectroscopy, infrared spectroscopy, Raman spectroscopy, nuclear magnetic resonance spectroscopy, photoemission spectroscopy, Mössbauer spectroscopy and so on.

Mass Spectrometry

Mass spectrometry measures mass-to-charge ratio of molecules using electric and magnetic fields. There are several ionization methods: electron impact, chemical ionization, electrospray, fast atom bombardment, matrix assisted laser desorption ionization, and others. Also, mass spectrometry is categorized by approaches of mass analyzers: magnetic-sector,quadrupole mass analyzer, quadrupole ion trap, Time-of-flight, Fourier transform ion cyclotron resonance, and so on.

Crystallography

Crystallography is a technique that characterizes the chemical structure of materials at the atomic level by analyzing the diffraction patterns of usually x-rays that have been deflected by atoms in the material. From the raw data the relative placement of atoms in space may be determined.

Electrochemical Analysis

Electroanalytical methods measure the potential (volts) and/or current (amps) in an electrochemical cell containing the analyte These methods can be categorized according to which aspects of the cell are controlled and which are measured. The three main categories are potentiometry (the difference in electrode potentials is measured), coulometry (the cell's current is measured over time), and voltammetry (the cell's current is measured while actively altering the cell's potential).

Thermal Analysis

Calorimetry and thermogravimetric analysis measure the interaction of a material and heat.

Separation

Separation processes are used to decrease the complexity of material mixtures. Chromatography and electrophoresis are representative of this field.

Hybrid Techniques

Combinations of the above techniques produce "hybrid" or "hyphenated" techniques. Several examples are in popular use today and new hybrid techniques are under development. For example, Gas chromatography-mass spectrometry, LC-MS, GC-IR, LC-NMR, LC-IR, CE-MS, and so on.

Hyphenated separation techniques refers to a combination of two (or more) techniques to detect and separate chemicals from solutions. Most often the other technique is some form of chromatography. Hyphenated techniques are widely used in chemistry and biochemistry. A slash is sometimes used instead of hyphen, especially if the name of one of the methods contains a hyphen itself.

Microscopy

The visualization of single molecules, single cells, biological tissues and nano-micro-materials is very important and attractive approach in analytical science. Also, hybridization with other traditional analytical tools is revolutionizing analytical science. Microscopy can be categorized into three different fields: optical microscopy, electron microscopy, and scanning probe microscopy. Recently, this field is rapidly progressing because of the rapid development of computer and camera industries.

Lab-on-a-chip

Devices that integrate (multiple) laboratory functions on a single chip of only millimeters to a few square centimeters in size and that are capable of handling extremely small fluid volumes down to less than pico liters.

METHODS AND DATA ANALYSIS

A standard method for analysis of concentration involves the creation of a calibration curve. This allows for determination of the amount of a chemical in a material by comparing the results of unknown sample to those of a series known standards.If the concentration of element or compound in a sample is too high for the detection range of the technique, it can simply be diluted in a pure solvent. If the amount in the sample is below an instrument's range of measurement, the method of addition can be used. In this method a known quantity of the element or compound under study is added, and the difference between the concentration added, and the concentration observed is the amount actually in the sample.

Internal Standards

Sometimes an internal standard is added at a known concentration directly to an analytical sample to aid in quantitation. The amount of analyte present is then determined relative to the internal standard as a calibrant.

APPLICATIONS

Analytical chemistry research is largely driven by performance (sensitivity, selectivity, robustness, linear range, accuracy, precision, and speed), and cost (purchase, operation, training, time, and space). Among the main branches of contemporary analytical atomic spectrometry, the most widespread and universal are optical and mass spectrometry (see Prospects in Analytical Atomic Spectrometry). In the direct elemental analysis of solid samples, the new leaders are laser-induced breakdown and laser ablation mass spectrometry, and the related techniques with transfer of the laser ablation products into inductively coupled plasma. Advances in design of diode lasers and optical parametric oscillators promote developments in fluorescence and ionization spectrometry and also in absorption techniques where uses of optical cavities for increased effective absorption pathlength are expected to expand. Steady progress and growth in applications of plasma- and laser-based methods are noticeable. An interest towards the absolute (standardless) analysis has revived, particularly in the emission spectrometry.

A lot of effort is put in shrinking the analysis techniques to chip size. Although there are few examples of such systems competitive with traditional analysis techniques, potential advantages include size/portability, speed, and cost. (micro Total Analysis System (µTAS) or Lab-on-a-chip). Microscale chemistry reduces the amounts of chemicals used.

Much effort is also put into analyzing biological systems. Examples of rapidly expanding fields in this area are:

- *Genomics* - DNA sequencing and its related research. Genetic fingerprinting and DNA microarray are very popular tools and research fields.
- *Proteomics* - the analysis of protein concentrations and modifications, especially in response to various stressors, at various developmental stages, or in various parts of the body.
- *Metabolomics* - similar to proteomics, but dealing with metabolites.
- *Transcriptomics* - mRNA and its associated field.
- *Lipidomics* - lipids and its associated field.

- *Peptidomics* - peptides and its associated field.
- *Metalomics* - similar to proteomics and metabolomics, but dealing with metal concentrations and especially with their binding to proteins and other molecules.

Analytical chemistry has played critical roles in the understanding of basic science to a variety of practical applications, such as biomedical applications, environmental monitoring, quality control of industrial manufacturing, forensic science and so on.

The recent developments of computer automation and information technologies have innervated analytical chemistry to initiate a number of new biological fields. For example, automated DNA sequencing machines were the basis to complete human genome projects leading to the birth of genomics. Protein identification and peptide sequencing by mass spectrometry opened a new field of proteomics. Furthermore, a number of ~omics based on analytical chemistry have become important areas in modern biology.

Also, analytical chemistry has been an indispensable area in the development of nanotechnology. Surface characterization instruments, electron microscopes and scanning probe microscopes enables scientists to visualize atomic structures with chemical characterizations.

Analytical chemistry is pursuing the development of practical applications and commercial instruments rather than elucidating scientific fundamentals. This may be an arguable difference from overlapping science areas such as physical chemistry and biophysics, although there isn't any distinct boundaries among disciplines in contemporary science and technology. However, this aspect may attract many engineers' interest; thus, it is not difficult to see papers from engineering departments in analytical chemistry journals.

Among active contemporary analytical chemistry research fields, micro total analysis system is considered as a great promise of revolutionary technology. In this approach, integrated and miniaturized analytical systems are being developed to control and analyze single cells and single molecules. This cutting-edge technology has a promising potential of leading a new revolution in science as integrated circuits did in computer developments.

CHAPTER - 2

Qualitative Inorganic Analysis

Classical qualitative inorganic analysis is a method of analytical chemistry which seeks to find elemental composition of inorganic compounds. It is mainly focused on detecting ions in an aqueous solution, so that materials in other forms may need to be brought into this state before using standard methods. The solution is then treated with various reagents to test for reactions characteristic of certain ions, which may cause color change, solid forming and other obviously visible changes.

DETECTING CATIONS

According to their properties, cations are usually classified into five groups. Each group has a common reagent which can be used to separate them from the solution. The separation must be done in the sequence specified below, otherwise, for example, some ions of 1st group can also react with 2nd group reagent, so that the solution must not have any ions left from previous groups to obtain meaningful results. The division and precise details of separating into groups vary slightly from one source to another; given below is one of the commonly used schemes.

First Analytical Group of Cations

First analytical group of cations consists of ions that form insoluble chlorides. As such, the group reagent to separate them is hydrochloric acid, usually used at a concentration of 1–2 M. Concentrated HCl must not be used, because it forms a soluble complex ion ($[PbCl_4]^{2-}$) with Pb^{2+}. Consequently the Pb^{2+} ion would go undetected.

The most important cations in 1st group are Ag^+, Hg_2^{2+}, and Pb^{2+}. The chlorides of these elements cannot be distinguished from each other by their colour - they are all white solid compounds. $PbCl_2$ is soluble in hot water, and can therefore be differentiated easily. Ammonia is used as a reagent to distinguish between the other two. While AgCl dissolves in ammonia (due to the formation of the complex ion $[Ag(NH_3)_2]^+$), Hg_2Cl_2 gives a black precipitate consisting of a mixture of chloro-mercuric amide and elemental mercury. Furthermore, AgCl is reduced to silver under light, which gives samples a violet colour.

$PbCl_2$ is far more soluble than the chlorides of the other two ions, especially in hot water. Therefore, HCl in concentrations which completely precipitate Hg_2^{2+} and Ag^+, may not be sufficient to do the same to Pb^{2+}. Higher concentrations of Cl^- cannot be used for the aforementioned reasons. Thus, a filtrate obtained after first group analysis of Pb^{2+} contains an appreciable concentration of this cation, enough to give the test of the second group, viz. formation of an insoluble sulfide. For this reason, Pb^{2+} is usually also included in the 2nd analytical group.

Second Analytical Group of Cations

The *second analytical group of cations* consists of ions that forms acid-insoluble sulfides. Cations in the 2nd group include: Cd^{2+}, Bi^{3+}, Cu^{2+}, As^{3+}, As^{5+}, Sb^{3+}, Sb^{5+}, Sn^{2+}, Sn^{4+} and Hg^{2+}. Pb^{2+} is usually also included here in addition to the first group.

The reagent can be any substance that gives S^{2-} ions in such solutions; most commonly used are H_2S (at 0.2-0.3 M), AKT (at 0.3-0.6 M). The test with the sulfide ion must be conducted in the presence of dilute HCl. Its purpose is to keep the sulfide ion concentration at a required minimum, so as to allow the precipitation of 2nd group cations alone. If dilute acid is not used, the early precipitation of 4th group cations (if present in solution) may occur, thus leading to misleading results. Acids beside HCl are rarely used. Sulfuric acid may lead to the precipitation of the 4th group cations, while nitric acid directly reacts with the sulfide ion (reagent), forming colloidal sulfur.

The precipitates of these cations are almost indistinguishable, except for CdS which is yellow. All the precipitates, except for HgS, are soluble in dilute mineral acids. HgS is soluble only in aqua regia, which can be used to separate it from the rest. The action of ammonia is also useful in differentiating the cations. CuS dissolves in ammonia forming an intense blue solution, while CdS dissolves forming a colourless solution. The sulfides of As^{3+}, As^{5+}, Sb3+, Sb5+, Sn2+, Sn4+ are soluble in yellow ammonium sulfide, where they form polysulfide complexes.

Third Analytical Group of Cations

The third analytical group of cations includes ions that form sulfides which are insoluble in basic solution. The reagents are similar to these of the 2nd group, but separation is conducted at pH of 8-9. Occasionally, a buffer solution is used to ensure this pH.

Cations in the 3rd group are, among others: Zn^{2+}, Ni^{2+}, Co^{2+}, Mn^{2+}, Fe^{2+}, Fe^{3+}, Al^{3+}, and Cr^{3+}.

Fourth Analytical Group of Cations

Ions in *the fourth analytical group of cations* form carbonates that are insoluble in water. The reagent usually used is $(NH_4)_2CO_3$ (at around 0.2 M), the pH should be neutral or slightly basic.

Caution should be taken to properly separate all lower analytical groups beforehand, as many of cations in previous groups also form insoluble carbonates.

Most important ions in the 4th group: Ba^{2+}, Ca^{2+}, and Sr^{2+}. After separation, the easiest way to distinguish these ions is by testing flame colour: barium gives a yellow-green flame, calcium orange-red and strontium deep red.

Fifth Analytical Group of Cations

Cations which are left after carefully separating previous groups are considered to be in the *fifth analytical group*. The most important ones are Mg^{2+}, Li^{+}, Na^{+}, K^{+} and NH^{4+}.

DETECTING ANIONS

Halides are precipitated by silver nitrate; they can be further identified by color. Sulfates can be precipitated by barium chloride. Nitrates can be reduced to ammonia.

First Analytical Group of Anions

The *first group of anions* consist of CO_3^{2-}, HCO_{3-}, CH_3COO^-, S^{2-}, SO_3^{2-}, $S_2O_3^{2-}$ and NO_2^-. The group reagent for Group 1 anion is dilute hydrochloric acid (HCL) or dilutes sulfuric acid H_2SO_4.

- Carbonates give a brisk effervescence with dilute H_2SO_4 due to the release of Co_2, a colourless gas which turns lime water milky due to formation of $CaCO_3$. The milkiness disappears on passing the lime water through an excess of the gas, due to formation of $Ca(HCO)_3$.
- Acetates give the vinegar-like smell of CH_3COOH when treated with dilute H_2SO_4. A blood red colouration is produced upon addition of yellow FeCl3, due to formation of iron(III) acetate.

- Sulphides give the rotten egg smell of H_2S when treated with dilute H_2SO_4. The presence of sulfide is confirmed by adding lead(II) acetate paper, which turns black due to the formation of PbS. Sulfides also turn solutions of red solution nitroprusside purple.
- Sulfites produce SO_2 gas, which smells of burning sulfur, when treated with dilute acid. They turn acidified $K_2Cr_2O_7$ from orange to green.
- Nitrites give reddish brown fumes of NO_2 when treated with dilute H_2SO_4. These fumes cause a solution of potassium iodide (KI) and starch to turn blue.

Second Analytical Group of Anions

The *second group of anions* consists of Cl^-, Br^-, I^-, NO^-_3 and $C_2O_2{}^-{}_4$. The group reagent for Group 2 anion is concentrated sulfuric acid (H_2SO_4).

After addition of the acid, chlorides, bromides and iodides will form precipitates with silver nitrate. The precipitates are white, pale yellow and yellow, respectively. The silver halides formed are completely soluble, partially soluble, or not soluble at all, respectively, in ammonium hydroxide (NH_4OH) solution.

Chlorides are confirmed by the *chromyl chloride test.* When the salt is heated with $K_2Cr_2O_7$ and concentrated H_2SO_4, red vapours of chromyl chloride (CrO_2Cl_2) are produced. Passing this gas through a solution of NaOH produces a yellow solution of Na_2CrO_4. The acidified solution of Na_2CrO_4 gives a yellow precipitate with the addition of $(CH_3COO)_2Pb$.

Bromides and iodides are confirmed by the *layer test.* A sodium carbonate extract is made from the solution containing bromide or iodide, and $CHCL_3$ or CS_2 is added to the solution, which separates into two layers: a brown colour in the $CHCl_3$ or CS_2 layer indicates the presence of Br^-, and a violet colour indicates the presence of I^-.

Nitrates give brown fumes with concentrated H_2SO_4 due to formation of NO_2. This is intensified upon adding copper turnings. Nitrate ion is confirmed by adding an aqueous solution of the salt to $FeSO_4$ and pouring concentrated H_2SO_4 slowly along the sides of the test tube, which produces a brown ring around the walls of the tube, caused by the formation of $Fe(NO)^{2+}$[1].

Upon treatment with concentrated sulfuric acid, oxalates yield colourless CO_2 and CO gases. These gases burn with a blue flame and turn lime water milky. Oxalates also decolourise $KMnO_4$ and give a white precipitate with $CaCl_2$.

MODERN TECHNIQUES

Qualitative inorganic analysis is now used only as a pedagogical tool. Modern techniques such as atomic absorption spectroscopy and ICP-MS are able to quickly detect the presence and concentrations of elements using a very small amount of sample.

QUANTITATIVE ANALYSIS (CHEMISTRY)

In chemistry, quantitative analysis is the determination of the absolute or relative abundance (often expressed as a concentration) of one, several or all particular substance(s) present in a sample.

Once the presence of certain substance(s) in a sample is known, the study of their absolute or relative abundance can help in determining specific properties. Knowing the composition of a sample is very important and several ways have been developed to make it possible, like gravimetric and volumetric analysis. Gravimetric analysis yields more accurate data about the composition of a sample than volumetric analysis does, but the first one takes more time to perform in the laboratory and is a little boring. Volumetric analysis in the other side doesn't take that much time and the results that we obtain are in the most cases satisfactory. Volumetric analysis can be simply a titration based in a neutralization reaction but it can also be a precipitation or a complex forming reaction as well as a titration based in an redox reaction. It is important to note, however, that each method in quantitative analysis has a general specification, in neutralization reactions, for example, the reaction that occurs is between an acid and a base, which yields a salt and water, hence the name neutralization. In the precipitation reactions the standard solution is in the most cases silver nitrate which is used as a reagent to react with the ions present in the sample and to form a high insoluble precipitate. Precipitation methods are often called simply as Argentometry. In the two other methods the situation is the same. Complex forming titration is a reaction that occurs between metal ions and a standard solution that is in the most cases EDTA (ethylene diamine tetra acetic acid). In the redox titration that reaction is carried out between an oxidizing agent and a reduction agent.

For example, quantitative analysis performed by mass spectrometry on biological samples can determine, by the relative abundance ratio of specific proteins, indications of certain diseases, like cancer.

QUANTITATIVE vs. QUALITATIVE

The term "quantitative analysis" is often used in comparison (or contrast) with "qualitative analysis", which seeks information about the identity or form of substance present. For instance, a chemist might be given an unknown solid sample. He or she will use "qualitative" techniques (perhaps NMR or IR spectroscopy) to identify the compounds present, and then quantitative techniques to determine the amount of each compound in the sample. Careful procedures for recognizing the presence of different metal ions have been developed, although they have largely been replaced by modern instruments; these are collectively known as qualitative inorganic analysis. Similar tests for identifying organic compounds (by testing for different functional groups) are also known.

Many techniques can be used for either qualitative or quantitative measurements. For instance, suppose an indicator solution changes color in the presence of a metal ion. It could be used as a qualitative test: does the indicator solution change color when a drop of sample is added? It could also be used as a quantitative test, by studying the colour of the indicator solution with different concentrations of the metal ion. (This would probably be done using ultraviolet-visible spectroscopy.)

LIST OF CHEMICAL ANALYSIS METHODS

A

Atomic absorption spectroscopy (AAS)

Atomic emission spectroscopy (AES)

Atomic fluorescence spectroscopy (AFS)

Alpha particle X-ray spectrometer (APXS)

C

Capillary electrophoresis (CE)

Chromatography

Colorimetry

Computed tomography

Cyclic Voltammetry (CV)

D

Differential scanning calorimetry (DSC)

E

Electron paramagnetic resonance (EPR) also called Electron spin resonance (ESR)

Energy Dispersive Spectroscopy (EDS/EDX)

F

Field Flow Fractionation (FFF)

Flow Injection Analysis (FIA)

Fourier transform spectroscopy (FTIR)

G

Gas chromatography (GC)

Gas chromatography-mass spectrometry (GC-MS)

H

High Performance Liquid Chromatography (HPLC)

I

Ion Microprobe (IM)

Inductively coupled plasma (ICP)

Ion selective electrode (ISE) eg. determination of pH

L

Laser Induced Breakdown Spectroscopy (LIBS)

M

Mass spectrometry (MS)

Mossbauer spectroscopy

N

Neutron activation analysis

Nuclear magnetic resonance (NMR)

P

Particle induced X-ray emission spectroscopy (PIXE)

Pyrolysis gas chromatography mass spectrometry (PY-GC-MS)

R

Raman Spectroscopy

Refractive index

Resonance enhanced multiphoton ionization (REMPI)

S

Scanning electron microscope (SEM)

Scanning X-ray Microscope (SXM) also called Scanning transmission X-ray Microscope (STXM)

T

Transmission electron microscopy (TEM)

X

X-ray diffraction (XRD)

X-ray fluorescence spectroscopy (XRF)

X-ray microscopy (XRM)

CHAPTER - 3

Ultracentrifuge

The ultracentrifuge is a centrifuge optimized for spinning a rotor at very high speeds, capable of generating acceleration as high as 1,000,000 g (9,800 km/s^2). There are two kinds of ultracentrifuges, the preparative and the analytical ultracentrifuge. Both classes of instruments find important uses in molecular biology, biochemistry and polymer science. Theodor Svedberg invented the analytical ultracentrifuge in 1923, and won the Nobel Prize in Chemistry in 1926 for his research on colloids and proteins using the ultracentrifuge.

The vacuum ultracentrifuge was invented by Edward Greydon Pickels. It was his contribution of the vacuum which allowed a reduction in friction generated at high speeds. Vacuum systems also enabled the maintenance of constant temperature.

In 1946, Pickels cofounded Spinco (Specialized Instruments Corp.) and marketed an ultracentrifuge based on his design. Pickels, however, considered his design to be complicated and developed a more "foolproof" version. But even with the enhanced design, sales of the technology remained low, and Spinco almost went bankrupt. The company survived and was the first to commercially manufacture ultracentrifuges, in 1947. In 1949, Spinco introduced the Model L, the first preparative ultracentrifuge to reach a maximum speed of 40,000 rpm. In 1954, Beckman Instruments (now Beckman Coulter) purchased the company, forming the basis of its Spinco centrifuge division.

Fig. 3.1: A Laboratory Ultracentrifuge

ANALYTICAL ULTRACENTRIFUGE

In an analytical ultracentrifuge, a sample being spun can be monitored in real time through an optical detection system, using ultraviolet light absorption and/or interference optical refractive index sensitive system. This allows the operator to observe the evolution of the sample concentration versus the axis of rotation profile as a result of the applied centrifugal field. With modern instrumentation, these observations are electronically digitized and stored for further mathematical analysis. Two kinds of experiments are commonly performed on these instruments: sedimentation velocity experiments and sedimentation equilibrium experiments.

Sedimentation velocity experiments aim to interpret the entire time-course of sedimentation, and report on the shape and molar mass of the dissolved macromolecules, as well as their size-distribution The size resolution of this method scales approximately with the square of the particle radii, and by adjusting the rotor speed of the experiment size-ranges from 100 Da to 10 GDa can be covered. Sedimentation velocity experiments can also be used to study reversible chemical equilibria between macromolecular species, by either monitoring the number and molar mass of macromolecular complexes, by gaining information about the complex composition from multi-signal analysis exploiting differences in each components spectroscopic signal, or by following the composition dependence of the sedimentation rates of the macromolecular system, as described in Gilbert-Jenkins theory.

Sedimentation equilibrium experiments are concerned only with the final steady-state of the experiment, where sedimentation is balanced by diffusion opposing the concentration gradients, resulting in a time-independent concentration profile. Sedimentation equilibrium distributions in the centrifugal field are characterized by Boltzmann distributions. This experiment is insensitive to the shape of the macromolecule, and directly reports on the molar mass of the macromolecules and, for chemically reacting mixtures, on chemical equilibrium constants.

The kinds of information that can be obtained from an analytical ultracentrifuge include the gross shape of macromolecules, the conformational changes in macromolecules, and size distributions of macromolecular samples. For macromolecules, such as proteins, that exist in chemical equilibrium with different non-covalent complexes, the number and subunit stoichiometry of the complexes and equilibrium constant constants can be studied.

PREPARATIVE ULTRACENTRIFUGE

Preparative ultracentrifuges are available with a wide variety of rotors suitable for a great range of experiments. Most rotors are designed to hold tubes that contain the samples. *Swinging bucket rotors* allow the tubes to hang on hinges so the tubes reorient to the horizontal as the rotor initially accelerates. *Fixed angle rotors* are made of a single block of metal and hold the tubes in cavities bored at a predetermined angle. *Zonal rotors* are designed to contain a large volume of sample in a single central cavity rather than in tubes. Some zonal rotors are capable of dynamic loading and unloading of samples while the rotor is spinning at high speed.

Preparative rotors are used in biology for pelleting of fine particulate fractions, such as cellular organelles (mitochondria, microsomes, ribosomes) and viruses. They can also be used for gradient separations, in which the tubes are filled from top to bottom with an increasing concentration of a dense substance in solution. Sucrose gradients are typically used for separation of cellular organelles. Gradients of caesium salts are used for separation of nucleic acids. After the sample has spun at high speed for sufficient time to produce the separation, the rotor is allowed to come to a smooth stop and the gradient is gently pumped out of each tube to isolate the separated components.

HAZARDS

The tremendous rotational kinetic energy of the rotor in an operating ultracentrifuge makes the catastrophic failure of a spinning rotor a serious concern. The stresses of routine use and harsh chemical solutions eventually cause rotors to deteriorate. Proper use of the instrument and rotors within recommended limits and careful maintenance of rotors to prevent corrosion and to detect deterioration are necessary to avoid this hazard.

ZIPPE-TYPE CENTRIFUGE

The Zippe-type centrifuge is a particular design of gas centrifuge. It was developed in the Soviet Union by a team of 60 German scientists working in detention, captured after World War II. The centrifuge is named after the team's lead experimenter, Gernot Zippe.

Natural uranium consists of three isotopes; the majority (99.274 per cent) is U-238, while approximately 0.72 per cent is U-235 and the remaining 0.0055 per cent is U-234. If natural uranium is enriched to contain 3 per cent U-235, it can be used as fuel for light water nuclear reactors. If it is enriched to contain 90 per cent Uranium-235, it can be used for nuclear weapons.

Enriching uranium is difficult because the isotopes are very similar in weight: U-235 is only 1.26% lighter than u-238. It requires a centrifuge that can spin at 1,500 revolutions per second (90,000 RPM). For comparison, automatic washing machines operate at only about 12 to 25 revolutions per second during the spin cycle.

The device has a hollow, cylindrical rotor filled with gaseous uranium in the form of its hexafluoride. A pulsating magnetic field at the bottom of the rotor, similar to that used in an electric motor, is able to spin it quickly enough that the U-238 is thrown towards the edge. The lighter U-235 collects in the center. The bottom of the gaseous mix is heated, producing convection currents that move the U-238 down. The U-235 moves up, where scoops collect it.

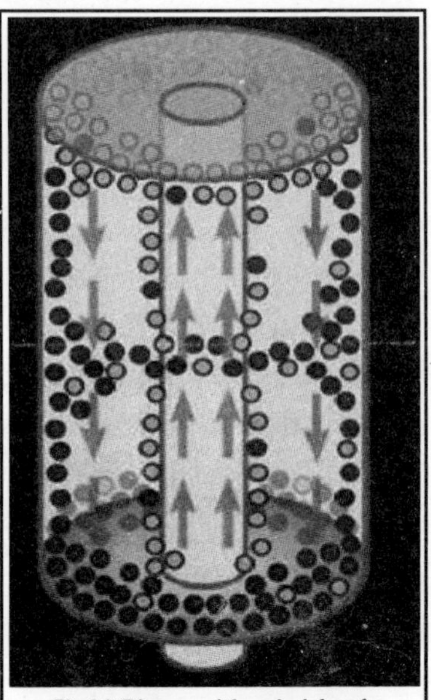

Fig. 3.2: **Diagram of the principles of a Zippe-type gas centrifuge with U-238 represented in dark blue and U-235 represented in light blue**

To reduce friction, the rotor spins in a vacuum. A magnetic bearing holds the top of the rotor steady, and the only physical contact is the needle-like bearing that the rotor sits on.

After the captured scientists were released in 1956, Gernot Zippe was surprised to find that engineers in the West were years behind in their centrifuge technology. He was able to reproduce his design at the University of Virginia in the United States, publishing the results, even though the Soviets had confiscated his notes. Dr. Zippe left the United States when he was effectively barred from continuing his research: The Americans classified the work as secret, requiring him to become an American citizen or return to Europe. He returned to Europe where, during the 1960s, he and his colleagues made the centrifuge more efficient by changing the material of the rotor from aluminum to a stronger alloy called maraging steel, which allowed it to spin even faster. This improved centrifuge design is used by the commercial company Urenco to produce enriched uranium fuel for nuclear power stations.

The exact details of advanced Zippe-type centrifuges are closely guarded secrets, but the efficiency of the centrifuges is improved by making them longer, and increasing their speed of rotation. To do so, even stronger materials, such as carbon fiber reinforced composite materials are used, and various techniques are used to avoid forces causing destructive vibrations, including the use of flexible "bellows" to allow controlled flexing of the rotor, as well as very careful control of the rotation speed to ensure that the centrifuge does not operate for very long at speeds where resonance is a problem.

The Zippe-type is difficult to build successfully, and it requires very carefully machined parts. To give some idea of the precision required, it was reported in 2006 that the tiny amount of material deposited in fingerprints on Iran's prototype centrifuges were enough to cause the machines to shatter However, compared to other enrichment methods, it is much cheaper and can be used in relative secrecy. This makes it ideal for covert nuclear-weapons programs and possibly increases the risk of nuclear proliferation. Centrifuge cascades also have much less material held up in the machine at any time, unlike gaseous diffusion plants.

GLOBAL USE

In 2004 Abdul Qadeer Khan, a Pakistani engineer, admitted to operating a smuggling ring responsible for supplying at least three countries with Zippe-type centrifuges.

Pakistan's nuclear program developed the P1 and P2 centrifuges—the first two centrifuges that Pakistan deployed in large numbers. The P1 centrifuge uses an aluminum rotor, and the P2 centrifuge uses a maraging steel rotor, which is stronger, spins faster, and therefore enriches more uranium per machine than the P1 centrifuge's aluminum rotor.

GAS CENTRIFUGE

A gas centrifuge is a separating machine specifically developed to separate Uranium-235 from Uranium-238. The gas centrifuge relies on the principles of centripetal force accelerating molecules based upon mass. When performed in a column, this yields separations of each component. The gas centrifuge was developed to replace the gaseous diffusion method of Uranium-235 extraction. The significant advantage of the gas centrifuge is that it relies on multiple centrifugal runs using cascades of centrifuges. This process yields higher concentrations of the Uranium-235 isotopes with significantly less energy usage compared to the previous gaseous diffusion process.

Centrifugal Process (How it works)

The centrifugal process relies on the effect of simulating gravity (using centripetal acceleration) to separate molecules according to their mass, and can be applied to both liquid and gaseous materials. Forces are applied by placing material inside a mechanism that rotates the material at high speed.

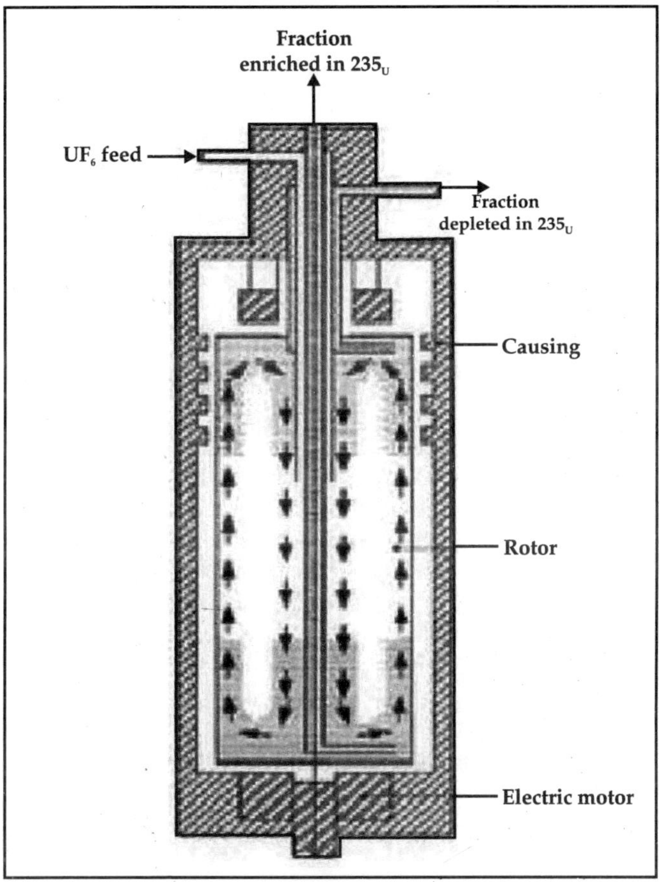

Fig. 3.3: Diagram of a gas centrifuge

Gas Centrifugation Process

The gas centrifugation process utilizes a unique design that allows gas to constantly flow in and out of the centrifuge. Unlike most centrifuges which rely on batch processing, the gas centrifuge utilizes continuous processing, allowing cascading, in which multiple identical processes occur in succession. The gas centrifuge consists of a cylindrical rotor, a casing, an electric motor, and three lines for material to travel. The gas centrifuge is designed with a casing that completely encloses the centrifuge. The cylindrical rotor is located inside the casing, which is evacuated of all air to produce a near frictionless rotation when operating. The motor spins the rotor, creating the centripetal force on the components as they enter the cylindrical rotor. There are two output lines, one located at the top of the centrifuge and the other located at the bottom. The heavier molecules will segregate to the bottom of the centrifuge while the lighter molecules will segregate to the top of the centrifuge. The output lines take these separations to other centrifuges to continue to the centrifugation process.

Separative Work Units (SWU)

The Separative Work Unit (SWU) is a measure of the amount of work done by the centrifuge and has units of mass (typically *kilogram separative work unit*). The work W_{SWU} necessary to separate a mass F of feed of assay x_f into a mass P of product assay x_p, and tails of mass T and assay x_t is expressed in terms of the number of separative work units needed, given by the expression.

$W_{SWU} = P.V(x_p) + T.V(x_t) - F.V(x_f)$ where $V(x)$ is the value function, defined as

$$V(x) = (1-2x)\,.\ln\left(\frac{1-x}{x}\right)$$

Practical application of centrifugation: *Separating Uranium-235 from Uranium-238*

The separation of uranium requires a gaseous form, Uranium hexafluoride (UF_6). Upon entering the centrifuge, the lighter Uranium-235 isotope collects towards one end of the centrifuge while the heavier Uranium-238 isotope at the other. The enriched uranium is upstream to another centrifuging stage and the depleted uranium downstream to the previous stage. Usually, enrichments plants may include thousands of centrifuges arranged in cascades.

HISTORY OF THE GAS CENTRIFUGE

Suggested in 1919, the centrifugal process was first successfully performed in 1934. J.W. Beams and co-workers at the University of Virginia developed the process by separating two chlorine isotopes through a vacuum ultracentrifuge. Although the Manhattan Project abandoned this theory during the development of the first nuclear weapons, the gas centrifuge uses much less energy than previous methods of separation The Nuclear program in the Soviet Union uses this advanced technology as the primary way to obtain enriched uranium since 1945.

CHAPTER - 4
Atomic Absorption Spectroscopy

In analytical chemistry, atomic absorption spectroscopy is a technique for determining the concentration of a particular metal element in a sample The technique can be used to analyze the concentration of over 70 different metals in a solution.

Although atomic absorption spectroscopy dates to the nineteenth century, the modern form was largely developed during the 1950s by a team of Australian chemists. They were led by Alan Walsh and worked at the CSIRO (Commonwealth Science and Industry Research Organisation) Division of Chemical Physics in Melbourne, Australia.

PRINCIPLES

The technique makes use of absorption spectrometry to assess the concentration of an analyte in a sample. It relies therefore heavily on Beer-Lambert law.

In short, the electrons of the atoms in the atomizer can be promoted to higher orbitals for an instant by absorbing a set quantity of energy (i.e. light of a given wavelength). This amount of energy (or wavelength) is specific to a particular electron transition in a particular element, and in general, each wavelength corresponds to only one element. This gives the technique its elemental selectivity.

Fig. 4.1: Atomic absorption spectroscopy

As the quantity of energy (the power) put into the flame is known, and the quantity remaining at the other side (at the detector) can be measured, it is possible, from Beer-Lambert law, to calculate how many of these transitions took place, and thus get a signal that is proportional to the concentration of the element being measured.

INSTRUMENTATION

In order to analyze a sample for its atomic constituents, it has to be atomized. The sample should then be illuminated by light. The light transmitted is finally measured by a detector. In order to reduce the effect of emission from the atomizer (e.g. the black body radiation) or the environment, a spectrometer is normally used between the atomizer and the detector.

Types of Atomizer

The technique typically makes use of a flame to atomize the sample but other atomizers such as a graphite furnace or plasmas, primarily inductively coupled plasmas, are also used.

When a flame is used it is laterally long (usually 10 cm) and not deep. The height of the flame above the burner head can be controlled by adjusting the flow of the fuel mixture. A beam of light passes through this flame at its longest axis (the lateral axis) and hits a detector.

Analysis of Liquids

A liquid sample is normally turned into an atomic gas in three steps:

Desolvation (*Drying*) – the liquid solvent is evaporated, and the dry sample remains

Vaporization (*Ashing*) – the solid sample vaporises to a gas

Atomization – the compounds making up the sample are broken into free atoms.

Radiation Sources

The radiation source chosen has a spectral width narrower than that of the atomic transitions.

Hollow Cathode Lamps

Hollow cathode lamps are the most common radiation source in atomic absorption spectroscopy. Inside the lamp, filled with argon or neon gas, is a cylindrical metal cathode containing the metal for excitation, and an anode. When a high voltage is applied across the anode and cathode, gas particles are ionized. As voltage is increased, gaseous ions acquire enough energy to eject metal atoms from the cathode. Some of these atoms are in an excited states and emit light with the frequency characteristic to the metal Many modern hollow cathode lamps are selective for several metals.

Diode Lasers

Atomic absorption spectroscopy can also be performed by lasers, primarily diode lasers because of their good properties for laser absorption spectrometry The technique is then either referred to as diode laser atomic absorption spectrometry (DLAAS or DLAS) or, since wavelength modulation most often is employed, wavelength modulation absorption spectrometry.

BACKGROUND CORRECTION METHODS

The narrow bandwidth of hollow cathode lamps make spectral overlap rare. That is, it is unlikely that an absorption line from one element will overlap with another. Molecular emission is much broader, so it is more likely that some molecular absorption band will overlap with an atomic line. This can result in artificially high absorption and an improperly high calculation for the concentration in the solution. Three methods are typically used to correct for this:

- *Zeeman correction* - A magnetic field is used to split the atomic line into two sidebands (see Zeeman effect). These sidebands are close enough to the original wavelength to still overlap with molecular bands, but are far enough not to overlap with the atomic bands. The absorption in the presence and absence of a magnetic field can be compared, the difference being the atomic absorption of interest.
- *Smith-Hieftje correction* (*invented by Stanley B. Smith and Gary M. Hieftje*) - The hollow cathode lamp is pulsed with high current, causing a larger atom population and self-absorption during the pulses. This self-absorption causes a broadening of the line and a reduction of the line intensity at the original wavelength.
- *Deuterium lamp correction* - In this case, a separate source (a deuterium lamp) with broad emission is used to measure the background emission. The use of a separate lamp makes this method the least accurate, but its relative simplicity (and the fact that it is the oldest of the three) makes it the most commonly used method.

ABSORPTION SPECTROSCOPY

Absorption spectroscopy refers to a range of techniques employing the interaction of electromagnetic radiation with matter. (Spectroscopy is a word that has come to denote an even wider variety of techniques used in physics and chemistry.) In absorption spectroscopy, the intensity of a beam of light measured before and after interaction with a sample is compared. When combined with the word spectroscopy, the words transmission and remission refer to the direction of travel of the beam measured after absorption to that before. The descriptions of the experimental arrangement usually assume that there is a unique direction of light incident upon the sample, and that a plane perpendicular to this direction passes through the sample. Light that is scattered from the sample toward a detector on the opposite side of the sample is said to be detected in transmission and treated according to the theory of transmission spectroscopy. Light that is scattered from the sample toward a detector on the same side of the sample is said to be detected in remission and it is this light that is the subject of remission spectroscopy. The remitted radiation may be composed of two kinds of radiation referred to as specular reflection (when the angle of reflection is equal to the angle of incidence) and diffuse reflection (at all other angles).

Another descriptor associated with absorption spectroscopy is the wavelength range of the radiation being used in the incident beam. Thus you will find references to infrared spectroscopy, near infrared spectroscopy, microwave spectroscopy; all of which are examples of absorption spectroscopy. On the other hand you will also find references to other wavelength ranges, such as x-ray spectroscopy, that usually denote an emission spectroscopy. This article deals primarily with UV-visible spectroscopy.

UV-visible spectroscopy refers to techniques where one measures how much light of a particular wavelength (color) is absorbed by a sample. Since color can often be correlated with the presence and or structure of a particular chemical, and since absorbance is often an easy and cheap measurement to make, absorbance spectroscopy is widely used for both qualitative (is a chemical present?) and quantitative (how much?) and structural (is it degraded?) work in a wide range of fields. For instance, DNA absorbs light in the UV range (which is partly why sunlight is dangerous) so the amount of DNA in a sample can be determined by measuring the absorbance of UV light.

The relation between the visible colour and the absorbance colour is complicated; a sample that appears red does not absorb in the red, but absorbs at OTHER wavelengths (colours) so that the light which passes through the sample is enriched in red.

The word "colour" is placed in quotes to indicate that absorbance spectroscopy deals not only with light in the visible range - photons with a wavelength of roughly 400 to 700 nanometers, but also with wavelengths that lie outside of the range of human vision (IR, UV, X-rays). However, the principles are quite similar for both visible and nonvisible light.

More technically absorption spectroscopy is based on the absorption of photons by one or more substances present in a sample, which can be a solid, liquid, or gas, and subsequent promotion of electron(s) from one energy level to another in that substance. Note that the sample can be a pure, homogeneous substance or a complex mixture. The wavelength at which the incident photon is absorbed is determined by the difference in the available energy levels of the different substances present in the sample; it is the selectivity of absorbance spectroscopy - the ability to generate photon (light) sources that are absorbed by only some of the components in a sample - that gives absorbance spectroscopy much of its utility. Typically, X-rays are used to reveal chemical composition, and near ultraviolet to near infrared wavelengths are used to distinguish the configurations of various isomers in detail. In absorption spectroscopy the absorbed photons are not re-emitted (as in fluorescence) rather, the energy that is transferred to the chemical compound upon absorbance of a photon is lost by non-radiative means, such as transfer of energy as heat to other molecules.

While the relative intensity of the absorption lines do not vary with concentration, at any given wavelength the measured absorbance [$-\log(I/I_0)$] has been shown to be proportional to the molar concentration of the absorbing species and the thickness of the sample the light passes through. This is known as the Beer-Lambert law. The plot of amount of radiation absorbed versus wavelength for a particular compound is referred to as the absorption spectrum. The normalized absorption spectrum is characteristic for a particular compound, does not change with varying concentration and is like the chemical "fingerprint" of the compound. At wavelengths corresponding to the resonant energy levels of the sample, some of the incident photons are absorbed, resulting in a drop in the measured transmission intensity and a corresponding dip in the spectrum. The absorption spectrum can be measured using a spectrometer and by knowing the shape of the spectrum ,the optical path length and the amount of radiation absorbed, one can determine the structure and concentration of the compound.

Visible light absorption spectra can be taken in anything that is visibly clear. Polystyrene, quartz glass, and borosilicate (Pyrex) cells, often called cuvettes, are the most commonly used. UV light is absorbed by most glasses and plastics, so quartz cells are used. The Si-O moieties in glasses and quartz, and the C-C moieties in plastics absorb infrared light. Therefore, infrared absorption spectra are typically carried out with a thin film of the sample held in place between sodium chloride sample plates. Other methods involve suspending the compound in a substance that does not absorb in the region of study. Mineral oil (Nujol) emulsions and potassium bromide glasses are perhaps the most common. NaCl and KBr, being ionic, do not have significant IR absorptions, and Nujol has a relatively uncomplicated IR spectrum.

SPECTROSCOPY AS AN ANALYTICAL TOOL

Often it is of interest to know not only the chemical composition of a given sample, but also the relative concentrations of the several compositing compounds. To do this, a scale, or calibration curve, must be constructed using several known concentrations for each compound of interest. The resulting plot of concentration vs. absorbance is fit either by hand or using appropriate curve-fitting software, yielding a mathematical formula to determine the concentration in the sample. Repeating this process for each compound in a sample gives a model of several absorption spectra added together to reproduce the observed absorption. In this way it is possible, for instance, to measure the chemical composition of comets without actually bringing samples back to Earth.

A simple example: a cyanide standard at 200 parts per million gives an absorbance with an arbitrary value of 1540. An unknown sample gives a value of 834. The math could be stated as: "if 200 gives you 1540, what gives you 834?" Since this is a linear relation and goes through the origin, the unknown is easily calculated to be 108 parts per million. Note the beauty of the ratio method in that it is not necessary to know the values of the governing coefficients, or chromophores, or the experimental cell length - it all divides out.

In practice, use of a calibration curve rather than a single point of comparison reduces uncertainty in the final measurement by excluding random interference (noise) in the preparation of the standards.

ULTRAVIOLET-VISIBLE SPECTROSCOPY

Ultraviolet-visible spectroscopy or ultraviolet-visible spectrophotometry (UV-Vis or UV/Vis) involves the spectroscopy of photons in the UV-visible region. This means it uses light in the visible and adjacent (near ultraviolet (UV) and near infrared (NIR)) ranges. The absorption in the visible ranges directly affects the color of the chemicals involved. In this region of the electromagnetic spectrum, molecules undergo electronic transitions. This technique is complementary to fluorescence spectroscopy, in that fluorescence deals with transitions from the excited state to the ground state, while absorption measures transitions from the ground state to the excited state.

APPLICATIONS

UV/Vis spectroscopy is routinely used in the quantitative determination of solutions of transition metal ions and highly conjugated organic compounds.

- Solutions of transition metal ions can be coloured (i.e., absorb visible light) because d electrons within the metal atoms can be excited from one electronic state to another. The colour of metal ion solutions is strongly affected by the presence of other species, such as certain anions or ligands. For instance, the colour of a dilute solution of copper sulfate is a very light blue; adding ammonia intensifies the colour and changes the wavelength of maximum absorption (λ_{max}).
- Organic compounds, especially those with a high degree of conjugation, also absorb light in the UV or visible regions of the electromagnetic spectrum. The solvents for these determinations are often water for water soluble compounds, or ethanol for organic-soluble compounds. (Organic solvents may have significant UV absorption; not all solvents are suitable for use in UV spectroscopy. Ethanol absorbs very weakly at most wavelengths.) Solvent polarity and pH can effect the

absorption spectrum of an organic compound. Tyrosine, for example, increases in absorption maxima and molar extinction coefficient when pH increases from 6 to 13 or when solvent polarity decreases.

- While charge transfer complexes also give rise to colours, the colours are often too intense to be used for quantitative measurement.

The Beer-Lambert law states that the absorbance of a solution is directly proportional to the solution's concentration. Thus UV/VIS spectroscopy can be used to determine the concentration of a solution. It is necessary to know how quickly the absorbance changes with concentration. This can be taken from references (tables of molar extinction coefficients), or more accurately, determined from a calibration curve.

A UV/Vis spectrophotometer may be used as a detector for HPLC. The presence of an analyte gives a response which can be assumed to be proportional to the concentration. For accurate results, the instrument's response to the analyte in the unknown should be compared with the response to a standard; this is very similar to the use of calibration curves. The response (e.g., peak height) for a particular concentration is known as the response factor.

BEER-LAMBERT LAW

Main article: Beer-Lambert law

The method is most often used in a quantitative way to determine concentrations of an absorbing species in solution, using the Beer-Lambert law:

$$A = \log_{10} (I/I_0) = \in . c . L$$

where A is the measured absorbance, I_0 is the intensity of the incident light at a given wavelength, I is the transmitted intensity, L the pathlength through the sample, and c the concentration of the absorbing species. For each species and wavelength, e is a constant known as the molar absorptivity or extinction coefficient. This constant is a fundamental molecular property in a given solvent, at a particular temperature and pressure, and has units of $1/M^{*}$ *cm* or often AU/M^{*} *cm*.

The absorbance and extinction *e* are sometimes defined in terms of the natural logarithm instead of the base-10 logarithm.

The Beer-Lambert Law is useful for characterizing many compounds but does not hold as a universal relationship for the concentration and absorption of all substances. A 2nd order polynomial relationship between absorption and concentration is sometimes encountered for very large, complex molecules such as organic dyes (Xylenol Orange or Neutral Red, for example).

Practical Considerations

To actually make a valid measurement you must understand and be aware of the limitations of the particular instrument being used. This is *especially* important when making measurements using simple (and therefore relatively inexpensive) instruments, where a user is more likely to encounter an instrumental limitation, or when making measurements of materials that have not been well characterized yet.

The molar extinction coefficient, *e*, is a function of the wavelength (that is, the color) of the light used. For the Beer-Lambert relation above to hold in a particular case, the light must be *sufficiently* monochromatic that the exctinction coefficient used is well defined.

There can also be limits imposed by the materials being measured, for instance if the material is not a simple solution.

Notice that the Beer-Lambert law implies that changes in concentration and path length should have equivalent effects. That is, for example, diluting a solution by a factor of 10 should have the same effect on absorbance as shortening the path length from the normal 10 mm to 1 mm. If cells of different path length are available, this is an alternative test to simply plotting absorption versus concentration in order to judge the validity of a measurement. Failure of such a test could indicate a concentration-dependent effect in the sample, such as absorption flattening.

Spectral Bandwidth

For instance, the *spectral bandwidth* of the instrument (as FWHM), the portion of the spectrum selected for the measurement, must be much smaller than the width of the absorbance curve of the sample, so that the exctintion coefficient does not change significantly over the band. Some instruments allow selection of bandwidth. (The tradeoff is that reducing the bandwidth reduces the energy passed to the detector and will require a longer measurement time to achieve the same signal to noise ratio.)

Wavelength Error

In liquids, the extinction coefficient usually changes slowly with wavelength. A peak of the absorbance curve (a wavelength where the absorbance reaches a maximum) is where the rate of change in absorbance with wavelength is smallest. Measurements are usually made at a peak in order to minimize errors produced by errors in wavelength in the instrument, that is errors due to having an different extinction coefficient than assumed.

Stray Light

Another important factor is the *purity* of the light used. The most important factor affecting this is the *stray light* level of the monochromator. The detector used is broadband, it responds to all the light that reaches it. If a significant amount of the light passed through the sample contains wavelengths that have much lower extinction coefficients than the nominal one, the instrument will report an incorrectly low absorbance. Any instrument will reach a point where an increase in sample concentration will not result in an increase in the reported absorbance, because the detector is simply responding to the stray light. In practice the concentration of the sample or the optical path length must be adjusted to place the unknown absorbance within a range that is valid for the instrument. Sometimes an empirical calibration function is developed, using known concentrations of the sample, to allow measurements into the region where the instrument is becoming non-linear.

As a rough guide, an instrument with a single monochromator would typically have a stray light level corresponding to about 3 AU, which would make measurements above about 2 AU problematic. A more complex instrument with a double monochromator would have a stray light level corresponding to about 6 AU, which would therefore allow measuring a much wider absorbance range.

Absorption Flattening

The nature of the material being measured is also important. Solutions that are not homogeneous can show deviations from the Beer-Lambert law because of the phenomenon of absorption flattening. This can happen, for instance, where the absorbing substance is located within suspended particles. The deviations will be most noticeable under conditions of low concentration and high absorbance. The reference describes a way to correct for this deviation.

ULTRAVIOLET-VISIBLE SPECTROPHOTOMETER

The instrument used in ultraviolet-visible spectroscopy is called a UV/vis spectrophotometer. It measures the intensity of light passing through a sample (I), and compares it to the intensity of light before it passes through the sample (I_o). The ratio I/I_o is called the *transmittance,* and is usually expressed as a percentage (%T). The absorbance, *A,* is based on the transmittance:

$$A = -\log(\%T/100\%)$$

The basic parts of a spectrophotometer are a light source, a holder for the sample, a diffraction grating or monochromator to separate the different wavelengths of light, and a detector. The radiation source is often a Tungsten filament (300-2500 nm), a deuterium arc lamp which is continuous over the ultraviolet region (190-400 nm), and more recently light emitting diodes (LED) and Xenon Arc Lamps for the visible wavelengths. The detector is typically a photodiode or a CCD. Photodiodes are used with monochromators, which filter the light so that only light of a single wavelength reaches the detector. Diffraction gratings are used with CCDs, which collects light of different wavelengths on different pixels.

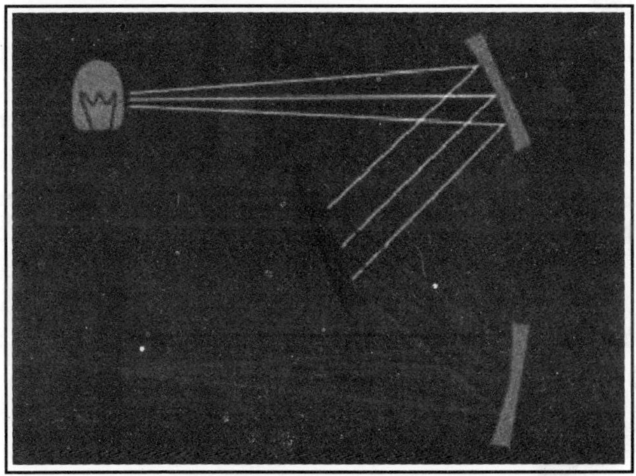

Fig. 4.2: Diagram of a single-beam UV/vis spectrophotometer

A spectrophotometer can be either single beam or double beam. In a single beam instrument (such as the

Spectronic 20), all of the light passes through the sample cell. I_o must be measured by removing the sample. This was the earliest design, but is still in common use in both teaching and industrial labs.

In a double-beam instrument, the light is split into two beams before it reaches the sample. One beam is used as the reference; the other beam passes through the sample. Some double-beam instruments have two detectors (photodiodes), and the sample and reference beam are measured at the same time. In other instruments, the two beams pass through a beam chopper, which blocks one beam at a time. The detector alternates between measuring the sample beam and the reference beam in synchronism with the chopper.

Samples for UV/Vis spectrophotometry are most often liquids, although the absorbance of gases and even of solids can also be measured. Samples are typically placed in a transparent cell, known as a cuvette. Cuvettes are typically rectangular in shape, commonly with an internal width of 1 cm. (This width becomes the path length, L, in the Beer-Lambert law.) Test tubes can also be used as cuvettes in some instruments. The type of sample container used must allow radiation to pass over the spectral region of interest. The most widely applicable cuvettes are made of high quality fused silica or quartz glass because these are transparent throughout the UV, visible and near infrared regions. Glass and plastic cuvettes are also common, although glass and most plastics absorb in the UV, which limits their usefulness to visible wavelengths.

CHAPTER - 5

Atomic Force Microscope

INTRODUCTION

The atomic force microscope (AFM) or scanning force microscope (SFM) is a very high-resolution type of scanning probe microscope, with demonstrated resolution of fractions of a nanometer, more than 1000 times better than the optical diffraction limit. The precursor to the AFM, the scanning tunneling microscope, was developed by Gerd Binnig and Heinrich Rohrer in the early 1980s, a development that earned them the Nobel Prize for Physics in 1986. Binnig, Quate and Gerber invented the first AFM in 1986. The AFM is one of the foremost tools for imaging, measuring and manipulating matter at the nanoscale. The information is gathered by "feeling" the surface with a mechanical probe. Piezoelectric elements that facilitate tiny but accurate and precise movements on (electronic) command enable the very precise scanning.

BASIC PRINCIPLE

The AFM consists of a microscale cantilever with a sharp tip (probe) at its end that is used to scan the specimen surface. The cantilever is typically silicon or silicon nitride with a tip radius of curvature on the order of nanometers.

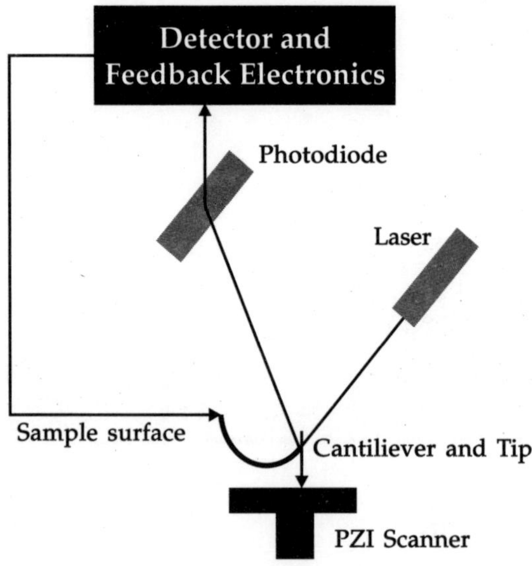

Fig 5.1: Diagram of Atomic Force Microscope

When the tip is brought into proximity of a sample surface, forces between the tip and the sample lead to a deflection of the cantilever according to Hooke's law. Depending on the situation, forces that are measured in AFM include mechanical contact force, Van der Waals forces, capillary forces, chemical bonding, electrostatic forces, magnetic forces, Casimir forces, solvation forces etc. As well as force, additional quantities may simultaneously be measured through the use of specialised types of probe (see Scanning thermal microscopy, photothermal microspectroscopy, etc.). Typically, the deflection is measured using a laser spot reflected from the top surface of the cantilever into an array of photodiodes. Other methods that are used include optical interferometry, capacitive sensing or piezoresistive AFM cantilevers. These cantilevers are fabricated with piezoresistive elements that act as a strain gauge. Using a Wheatstone bridge, strain in the AFM cantilever due to deflection can be measured, but this method is not as sensitive as laser deflection or interferometry.

If the tip was scanned at a constant height, a risk would exist that the tip collides with the surface, causing damage. Hence, in most cases a feedback mechanism is employed to adjust the tip-to-sample distance to maintain a constant force between the tip and the sample. Traditionally, the sample is mounted on a piezoelectric tube, that can move the sample in the z direction for maintaining a constant force, and the x and y directions for scanning the sample. Alternatively a 'tripod' configuration of three piezo crystals may be employed, with each responsible for scanning in the x,y and z directions. This eliminates some of the distortion effects seen with a tube scanner. In newer designs, the tip is mounted on a vertical piezo scanner while the sample is being scanned in X and Y using another piezo block. The resulting map of the area s = f(x, y) represents the topography of the sample.

The AFM can be operated in a number of modes, depending on the application. In general, possible imaging modes are divided into static (also called Contact) modes and a variety of dynamic (or non-contact) modes where the cantilever is vibrated.

IMAGING MODES

The primary modes of operation are static (contact) mode and dynamic mode. In the static mode operation, the static tip deflection is used as a feedback signal. Because the measurement of a static signal is prone to noise and drift, low stiffness cantilevers are used to boost the deflection signal. However, close to the surface of the sample, attractive forces can be quite strong, causing the tip to 'snap-in' to the surface. Thus static mode AFM is almost always done in contact where the overall force is repulsive. Consequently, this technique is typically called 'contact mode'. In contact mode, the force between the tip and the surface is kept constant during scanning by maintaining a constant deflection.

CONTACT MODE

In the dynamic mode, the cantilever is externally oscillated at or close to its fundamental resonance frequency or a harmonic. The oscillation amplitude, phase and resonance frequency are modified by tip-sample interaction forces; these changes in oscillation with respect to the external reference oscillation provide information about the sample's characteristics. Schemes for dynamic mode operation include frequency modulation and the more common amplitude modulation. In frequency modulation, changes in the oscillation frequency provide information about tip-sample interactions. Frequency can be measured with very high sensitivity and thus the frequency modulation mode allows for the use of very stiff cantilevers. Stiff cantilevers provide stability very close to the surface and, as a result, this technique was the first AFM technique to provide true atomic resolution in ultra-high vacuum conditions (Giessibl).

In amplitude modulation, changes in the oscillation amplitude or phase provide the feedback signal for imaging. In amplitude modulation, changes in the phase of oscillation can be used to discriminate between different types of materials on the surface. Amplitude modulation can be operated either in the non-contact or in the intermittent contact regime. In ambient conditions, most samples develop a liquid meniscus layer. Because of this, keeping the probe tip close enough to the sample for short-range forces to become detectable while preventing the tip from sticking to the surface presents a major hurdle for the non-contact dynamic mode in ambient conditions. Dynamic contact mode (also called intermittent contact or tapping mode) was developed to bypass this problem (Zhong et al.). In dynamic contact mode, the cantilever is oscillated such that the separation distance between the cantilever tip and the sample surface is modulated.

Amplitude modulation has also been used in the non-contact regime to image with atomic resolution by using very stiff cantilevers and small amplitudes in an ultra-high vacuum environment.

Tapping Mode

In tapping mode the cantilever is driven to oscillate up and down at near its resonance frequency by a small piezoelectric element mounted in the AFM tip holder. The amplitude of this oscillation is greater than 10 nm, typically 100 to 200 nm. Due to the interaction of forces acting on the cantilever when the tip comes close to the surface, Van der Waals force or dipole-dipole interaction, electrostatic forces, etc cause the amplitude of this oscillation to decrease as the tip gets closer to the sample. An electronic servo uses the piezoelectric actuator to control the height of the cantilever above the sample. The servo adjusts the height to maintain a set cantilever oscillation amplitude as the cantilever is scanned over the sample. A Tapping AFM image is therefore produced by imaging the force of the oscillating contacts of the tip with the sample surface. This is an improvement on conventional contact AFM, in which the cantilever just drags across the surface at constant force and can result in surface damage. Tapping mode is gentle enough even for the visualization of supported lipid bilayers or adsorbed single polymer molecules (for instance, 0.4 nm thick chains of synthetic polyelectrolytes) under liquid medium. At the application of proper scanning parameters, the conformation of single molecules remains unchanged for hours.

Non-Contact Mode

Here the tip of the cantilever does not contact the sample surface. Instead, during scanning, it oscillates above the adsorbed fluid layer on the surface. The cantilever is oscillated at a frequency slightly above its resonance frequency. The amplitude is typically a few nanometers (<10nm). The van der Waals forces, which extend from 1nm to 10nm above the adsorbed fluid layer, or any other long range force which extends above the surface decreases the resonance frequency of the cantilever. The decrease in resonance frquency causes the amplitude of oscillation to decrease. The feedback loop maintains a constant oscillation amplitude or frequency by vertically moving the scanner at each (x, y) data point until a "setpoint" amplitude or frequency is reached. The distance the scanner moves vertically at each (x, y) data point is stored by the computer to form the topographic image of the sample surface.

AFM Cantilever Deflection Measurement

Laser light from a solid state diode is reflected off the back of the cantilever and collected by a position sensitive detector (PSD) consisting of two closely spaced photodiodes whose output signal is collected by a differential amplifier. Angular displacement of cantilever results in one photodiode collecting more light than the other photodiode, producing an output signal (the difference between the photodiode signals normalized by their sum) which is proportional to the deflection of the cantilever. It detects cantilever deflections <1Å (thermal noise limited). A long beam path (several cm) amplifies changes in beam angle.

Force Spectroscopy

Another major application of AFM (besides imaging) is force spectroscopy, the measurement of force-distance curves. For this method, the AFM tip is extended towards and retracted from the surface as the static deflection of the cantilever is monitored as a function of piezoelectric displacement. These measurements have been used to measure nanoscale contacts, atomic bonding, Van der Waals forces, and Casimir forces, dissolution forces in liquids and single molecule stretching and rupture forces (Hinterdorfer & Dufrêne). Forces of the order of a few pico-Newton can now be routinely measured with a vertical distance resolution of better than 0.1 nanometer.

Problems with the technique include no direct measurement of the tip-sample separation and the common need for low stiffness cantilevers which tend to 'snap' to the surface. The snap-in can be reduced by measuring in liquids or by using stiffer cantilevers, but in the latter case a more sensitive deflection sensor is needed. By applying a small dither to the tip, the stiffness (force gradient) of the bond can be measured as well (Hoffmann et al.).

IDENTIFICATION OF INDIVIDUAL SURFACE ATOMS

The AFM can be used to image and manipulate atoms and structures on a variety of surfaces. The atom at the apex of the tip "senses" individual atoms on the underlying surface when it forms incipient chemical bonds with each atom. Because these chemical interactions subtly alter the tip's vibration frequency, they can be detected and mapped.

Physicist Oscar Custance (Osaka University, Graduate School of Engineering, Osaka, Japan) and his team used this principle to distinguish between atoms of silicon, tin and lead on an alloy surface.

The trick is to first measure these forces precisely for each type of atom expected in the sample. The team found that the tip interacted most strongly with silicon atoms, and interacted 23% and 41% less strongly with tin and lead atoms. Thus, each different type of atom can be identified in the matrix as the tip is moved across the surface.

Such a technique has been used now in biology and extended recently to cell biology. Forces corresponding to:

(i) the unbinding of receptor ligand couples;

(ii) unfolding of proteins;

(iii) cell adhesion at single cell scale have been gathered.

ADVANTAGES AND DISADVANTAGES

The AFM has several advantages over the scanning electron microscope (SEM). Unlike the electron microscope which provides a two-dimensional projection or a two-dimensional image of a sample, the AFM provides a true three-dimensional surface profile. Additionally, samples viewed by AFM do not require any special treatments (such as metal/carbon coatings) that would irreversibly change or damage the sample. While an electron microscope needs an expensive vacuum environment for proper operation, most AFM modes can work perfectly well in ambient air or even a liquid environment. This makes it possible to study biological macromolecules and even living organisms. In principle, AFM can provide higher resolution than SEM. It has been shown to give true atomic resolution in ultra-high vacuum (UHV) and, more recently, in liquid environments. High resolution AFM is comparable in resolution to Scanning Tunneling Microscopy and Transmission Electron Microscopy.

A disadvantage of AFM compared with the scanning electron microscope (SEM) is the image size. The SEM can image an area on the order of millimetres by millimetres with a depth of field on the order of millimetres. The AFM can only image a maximum height on the order of micrometres and a maximum scanning area of around 150 by 150 micrometres.

Another inconvenience is that an incorrect choice of tip for the required resolution can lead to image artifacts. Traditionally the AFM could not scan images as fast as an SEM, requiring several minutes for a typical scan, while a SEM is capable of scanning at near real-time (although at relatively low quality) after the chamber is evacuated. The relatively slow rate of scanning during AFM imaging often leads to thermal drift in the image, making the AFM microscope less suited for measuring accurate distances between artifacts on the image. However, several fast-acting designs were suggested to increase microscope scanning productivityincluding what is being termed videoAFM (reasonable quality images are being obtained with videoAFM at video rate - faster than the average SEM). To eliminate image distortions induced by thermodrift, several methods were also proposed.

AFM images can also be affected by hysteresis of the piezoelectric material (Lapshin, 1995) and cross-talk between the (x, y, z) axes that may require software enhancement and filtering. Such filtering could "flatten" out real topographical features. However, newer AFM use real-time correction software (for example, feature-oriented scanning, Lapshin, 2004, 2007) or closed-loop scanners which practically eliminate these problems. Some AFM also use separated orthogonal scanners (as opposed to a single tube) which also serve to eliminate cross-talk problems.

Due to the nature of AFM probes, they cannot normally measure steep walls or overhangs. Specially made cantilevers can be modulated sideways as well as up and down (as with dynamic contact and non-contact modes) to measure sidewalls, at the cost of more expensive cantilevers and additional artifacts.

PIEZOELECTRIC SCANNERS

AFM scanners are made from piezoelectric material, which expands and contracts proportionally to an applied voltage. Whether they elongate or contract depends upon the polarity of the voltage applied. The scanner is constructed by combining independently operated piezo electrodes for X, Y, & Z into a single tube, forming a scanner which can manipulate samples and probes with extreme precision in 3 dimensions.

Scanners are characterized by their sensitivity which is the ratio of piezo movement to piezo voltage, i.e. by how much the piezo material extends or contracts per applied volt. Because of differences in material or size, the sensitivity varies from scanner to scanner.

Sensitivity varies non-linearly with respect to scan size. Piezo scanners exhibit more sensitivity at the end than at the beginning of a scan. This causes the forward and reverse scans to behave differently and display hysteresis between the two scan directions. This can be corrected by applying a non-linear voltage to the piezo electrodes to cause linear scanner movement and calibrating the scanner accordingly.

The sensitivity of piezoelectric materials decreases exponentially with time. This causes most of the change in sensitivity to occur in the initial stages of the scanner's life. Piezoelectric scanners are run for approximately 48 hours before they are shipped from the factory so that they are past the point where we can expect large changes in sensitivity. As the scanner ages, the sensitivity will change less with time and the scanner would seldom require recalibration.

CHAPTER - 6

Chemical Substance

INTRODUCTION

A chemical substance is a material with a specific chemical composition. It is a concept that became firmly established in the late eighteenth century after work by the chemist Joseph Proust on the composition of some pure chemical compounds such as basic copper carbonate. He deduced that, "All samples of a compound have the same composition; that is, all samples have the same proportions, by mass, of the elements present in the compound." This is now known as the law of constant composition Later with the advancement of methods for chemical synthesis particularly in the realm of organic chemistry; the discovery of many more chemical elements and new techniques in the realm of analytical chemistry used for isolation and purification of elements and compounds from chemicals that led to the establishment of modern chemistry, the concept was defined as is found in most chemistry textbooks. However, there are some controversies regarding this definition mainly because the large number of chemical substances reported in chemistry literature need to be indexed.

A common example of a chemical substance is pure water; it has the same properties and the same ratio of hydrogen to oxygen whether it is isolated from a river or made in a laboratory. A pure chemical substance cannot be separated into other substances by a process that does not involve any chemical reaction and is rarely found in nature. Some typical chemical substances can be diamond, gold, salt (sodium chloride) and sugar (sucrose). Generally, chemical substances exist as a solid, liquid, or gas, and may change between these phases of matter with changes in temperature or pressure.

Forms of energy, such as light and heat, are not considered to be matter, and thus they are not "substances" in the scientific regard.

DEFINITION

Chemical substances (also sometimes referred to as a pure substances) are often defined as "any material with a definite chemical composition" in most introductory general chemistry textbooks. According to this definition a chemical substance can either be a pure chemical element or a pure chemical compound. However, there are exceptions to this definition; a pure substance can also be defined as a form of matter that has both definite composition and distinct properties The chemical substance index published by CAS also includes several alloys of uncertain composition. Non-stoichiometric compounds are a special case (in inorganic chemistry) that violates the law of constant composition, and for them, it is sometimes difficult to draw the line between a mixture and a compound, as in the case of palladium hydride.

CHEMICAL ELEMENTS

An element is a chemical substance that is made up of a particular kind of atoms and hence cannot be broken down or transformed by a chemical reaction into a different element, though it can be transmutated into another element through a nuclear reaction. This is so, because all of the atoms in a sample of an element have the same number of protons, though they may be different isotopes, with differing numbers of neutrons.

There are about 120 known elements, about 80 of which are stable - that is, they do not change by radioactive decay into other elements. However, the number of chemical substances that are elements can be more than 120, because some elements can occur as more than a single chemical substance (allotropes). For instance, oxygen exists as both diatomic oxygen (02) and ozone (03). The majority of elements are classified as metals. These are elements with a characteristic lustre such as iron, copper, and gold. Metals typically conduct electricity and heat well, and they are malleable and ductile Around a dozen elements, such as carbon, nitrogen, and oxygen, are classified as non-metals. Non-metals lack the metallic properties described above, they also have a high electronegativity and a tendency to form negative ions. Certain elements such as silicon sometimes resemble metals and sometimes resemble non-metals, and are known as metalloids.

CHEMICAL COMPOUNDS

A pure chemical compound is a chemical substance that is composed of a particular set of molecules or ions. Two or more elements combined into one substance, through a chemical reaction, form what is called a chemical compound. All compounds are substances, but not all substances are compounds.

A chemical compound can be either atoms bonded together in molecules or crystals in which atoms, molecules or ions form a crystalline lattice. Compounds based primarily on carbon and hydrogen atoms are called organic compounds, and all others are called inorganic compounds. Compounds containing bonds between carbon and a metal are called organometallic compounds.

Compounds in which components share electrons are known as covalent compounds. Compounds consisting of oppositely charged ions are known as ionic compounds, or salts.

In organic chemistry, there can be more than one chemical compound with the same composition and molecular weight. Generally, these are called isomers. Isomers usually have substantially different chemical properties, may be isolated and do not spontaneously convert to each other. A common example is glucose vs. fructose. The former is an aldehyde, the latter is a ketone. Their interconversion requires either enzymatic or acid-base catalysis. However, there are also tautomers, where isomerization occurs spontaneously, such that a pure substance cannot be isolated into its tautomers. A common example is glucose, which has open-chain and ring forms. One cannot manufacture pure open-chain glucose because glucose spontaneously cyclizes to the hemiacetal form.

SUBSTANCES VERSUS MIXTURES

All matter consists of various elements and chemical compounds, but these are often intimately mixed together. Mixtures contain more than one chemical substance, and they do not have a fixed composition. In principle, they can be separated into the component substances by purely mechanical processes. Butter, soil and wood are common examples of mixtures.

Grey iron metal and yellow sulfur are both chemical elements, and they can be mixed together in any ratio to form a yellow-grey mixture. No chemical process occurs, and the material can be identified as a mixture by the fact that the sulfur and the iron can be separated by a mechanical process, such as using a magnet to attract the iron away from the sulfur.

In contrast, if iron and sulfur are heated together in a certain ratio (56 grams (1 mol) of iron to 32 grams (1 mol) of sulfur), a chemical reaction takes place and a new substance is formed, the compound iron(II) sulfide, with chemical formula FeS. The resulting compound has all the properties of a chemical substance and is not a mixture. Iron(II) sulfide has its own distinct properties such as melting point and solubility, and the two elements cannot be separated using normal mechanical processes; a magnet will be unable to recover the iron, since there is no metallic iron present in the compound.

CHEMICALS VERSUS CHEMICAL SUBSTANCES

While the term *chemical substance* is a somewhat technical term used most often by professional chemists, the word *chemical* is more widely used in the pharmaceutical industry, government and society in general. Thus the word *chemical* includes a much wider class of substances that includes many mixtures of chemical substances that often find application in many vocations and is most commonly used only for artificial or processed substances, such as the products of the chemical industry.

NAMING AND INDEXING

Every chemical substance has one or more systematic names, usually named according to the IUPAC rules for naming. An alternative system is used by the Chemical Abstracts Service (CAS).

Many compounds are also known by their more common, simpler names, many of which predate the systematic name. For example, the long-known sugar glucose is now systematically named 6-(hydroxymethyl) oxane-2, 3, 4, 5-tetrol. Natural products and pharmaceuticals are also given simpler names, for example the mild pain-killer Naproxen is the more common name for the chemical compound (S)-6-methoxy-a-methyl-2-naphthaleneacetic acid.

Chemists frequently refer to chemical compounds using chemical formulae or molecular structure of the compound.

There has been a phenomenal growth in the number of chemical compounds being synthesized (or isolated), and then reported in the scientific literature by professional chemists around the world. An enormous number of chemical compounds are possible through the chemical combination of the known chemical elements. At the last count, about thirty million chemical compounds are known. The names of many of these compounds are often nontrivial and hence not very easy to remember or cite accurately. Also it is difficult to keep the track of them in the literature. Several international organizations like the IUPAC and the Chemical Abstract Service (CAS) have initiated steps to make such tasks easier. CAS that provides the abstracting services of the chemical literature, provides a numerical identifier, known as CAS registry number to each chemical substance that has been reported in the chemical literature (such as chemistry journals and patents). This information is compiled as a database and is popularly known as the Chemical substances index. Other computer-friendly systems that have been developed for substance information, are: SMILES and the International Chemical Identifier or InChI.

CHAPTER - 7

Forensic Science

Forensic science (often shortened to forensics) is the application of a broad spectrum of sciences to answer questions of interest to the legal system. This may be in relation to a crime or to a civil action. But besides its relevance to the underlying legal system, more generally forensics encompasses the accepted scholarly or scientific methodology and norms under which the facts regarding an event, or an artifact, or some other physical item (such as a corpse, or cadaver, for example) are to the broader notion of authentication whereby an interest outside of a legal form exists in determining whether an object is in fact what it purports to be, or is alleged as being.

The word "forensic" comes from the Latin adjective "*forensis*" meaning of or before the forum. During the time of the Romans, a criminal charge meant presenting the case before a group of public individuals in the forum. Both the person accused of the crime and the accuser would give speeches based on their side of the story. The individual with the best argument and delivery would determine the outcome of the case. Basically, the person with the sharpest forensic skills would win. This origin is the source of the two modern usages of the word "forensic" - as a form of legal evidence and as a category of public presentation.

In modern use, the term "forensics" in place of "forensic science" can be considered incorrect as the term "forensic" is effectively a synonym for "legal" or "related to courts". However, the term is now so closely associated with the scientific field that many dictionaries include the meaning that equates the word "forensics" with "forensic science".

HISTORY OF FORENSIC SCIENCE

The "Eureka" legend of Archimedes (287-212 BC) can be considered an early use of forensic anthropology. He determined that a crown was not completely made up of gold (as it was fraudulently claimed). This conclusion was reached by evaluating the density of the object using measurements of its displacement and its weight, as he was not allowed to damage the crown.The earliest account of fingerprint use to establish identity was during the 7th century. According to an Arabic merchant, Soleiman, a debtor's fingerprints were affixed to a bill, which would then be given to the lender. This bill was legally recognized as proof of the validity of the debt

The first written account of using medicine and entomology to solve (separate) criminal cases is attributed to the book *Xi Yuan Ji Lu* , translated as "Collected Cases of Injustice Rectified"), written in Song Dynasty China by Song 1186-1249) in 1248. In one of the accounts, the case of a person murdered with a sickle was solved by a death investigator who instructed everyone to bring his sickle to one location. Flies, attracted by the smell of blood, eventually gathered on a single sickle. In light of this, the murderer confessed. The book also offered advice on how to distinguish between a drowning (water in the lungs) and strangulation (broken neck cartilage), along with other evidence from examining corpses on determining if a death was caused by murder, suicide, or an accident.

In sixteenth century Europe, medical practitioners in army and university settings began to gather information on cause and manner of death. Ambroise Paré, a French army surgeon, systematically studied the effects of violent death on internal organs. Two Italian surgeons, Fortunato Fidelis and Paolo Zacchia, laid the foundation of modern pathology by studying changes which occurred in the structure of the body as the result of disease. In the late 1700s, writings on these topics began to appear. These included: "*A Treatise on Forensic Medicine and Public Health*" by the French physician Fodéré, and "*The Complete System of Police Medicine*" by the German medical expert Johann Peter Franck.

In 1776, Swedish chemist Carl Wilhelm Scheele devised a way of detecting arsenous oxide, simple arsenic, in corpses, although only in large quantities. This investigation was expanded, in 1806, by German chemist Valentin Ross, who learned to detect the poison in the walls of a victim's stomach, and by English chemist James Marsh, who used chemical processes to confirm arsenic as the cause of death in an 1836 murder trial.

Two early examples of English forensic science in individual legal proceedings demonstrate the increasing use of logic and procedure in criminal investigations. In 1784, in Lancaster, England, John Toms was tried and convicted for murdering Edward Culshaw with a pistol. When the dead body of Culshaw was examined, a pistol wad (crushed paper used to secure powder and balls in the muzzle) found in his head wound matched perfectly with a torn newspaper found in Toms' pocket. In Warwick, England, in 1816, a farm labourer was tried and convicted of the murder of a young maidservant. She had been drowned in a shallow pool and bore the marks of violent assault. The police found footprints and an impression from corduroy cloth with a sewn patch in the damp earth near the pool. There were also scattered grains of wheat and chaff. The breeches of a farm labourer who had been threshing wheat nearby were examined and corresponded exactly to the impression in the earth near the pool Later in the 20th century, several British pathologists, Bernard Spilsbury, Francis Camps, Sydney Smith and Keith Simpson would pioneer new forensic methods in Britain.

SUBDIVISIONS OF FORENSIC SCIENCE

- *Computational forensics* concerns the development of algorithms and software to assist forensic examination.
- *Criminalistics* is the application of various sciences to answer questions relating to examination and comparison of biological evidence, trace evidence, impression evidence (such as fingerprints, footwear impressions, and tire tracks), controlled substances, ballistics, firearm and toolmark examination, and other evidence in criminal investigations. Typically, evidence is processed in a crime lab.
- *Digital forensics* is the application of proven scientific methods and techniques in order to recover data from electronic/digital media. DF specialists work in the field as well as in the lab.
- *Forensic anthropology* is the application of physical anthropology in a legal setting, usually for the recovery and identification of skeletonized human remains.
- *Forensic archaeology* is the application of a combination of archaeological techniques and forensic science, typically in law enforcement.
- *Forensic DNA analysis* takes advantage of the uniqueness of an individual's DNA to answer forensic questions such as determining paternity/maternity or placing a suspect at a crime scene.
- *Forensic entomology* deals with the examination of insects in, on, and around human remains to assist in determination of time or location of death. It is also possible to determine if the body was moved after death.

- *Forensic geology* deals with trace evidence in the form of soils, minerals and petroleums.
- *Forensic interviewing* is a method of communicating designed to elicit information and evidence.
- *Forensic meteorology* is a site specific analysis of past weather conditions for a point of loss.
- *Forensic odontology* is the study of the uniqueness of dentition better known as the study of teeth.
- *Forensic pathology* is a field in which the principles of medicine and pathology are applied to determine a cause of death or injury in the context of a legal inquiry.
- *Forensic psychology* is the study of the mind of an individual, using forensic methods. Usually it determines the circumstances behind a criminal's behavior.
- *Forensic toxicology* is the study of the effect of drugs and poisons on/in the human body.
- *Forensic document examination* or questioned document examination answers questions about a disputed document using a variety of scientific processes and methods. Many examinations involve a comparison of the questioned document, or components of the document, to a set of known standards. The most common type of examination involves handwriting wherein the examiner tries to address concerns about potential authorship.
- *Veterinary Forensics* is forensics applied to crimes involving animals.

QUESTIONABLE FORENSIC TECHNIQUES

Some forensic techniques, believed to be scientifically sound at the time they were used, have turned out later to have much less scientific merit, or none. Some such techniques include:

- *Comparative bullet-lead analysis* was used by the FBI for over four decades, starting with the John F. Kennedy assassination in 1963. The theory was that each batch of ammunition possessed a chemical makeup so distinct that a bullet could be traced back to a particular batch, or even a specific box. However, internal studies and an outside study by the National Academy of Sciences found that the technique was unreliable, and the FBI abandoned the test in 2005.
- *Forensic dentistry has come under fire*; in at least two cases, bite mark evidence has been used to convict people of murder who were later freed by DNA evidence. A 1999 study by a member of the American Board of Forensic Odontology found a 63 per cent rate of false identifications and is commonly referenced within online news stories and conspiracy websites.· However, the study was based on an informal workshop during an ABFO meeting which many members did not consider a valid scientific setting.

LITIGATION SCIENCE

Litigation Science describes analysis or data developed or produced *expressly* for use in a trial, versus those produced in the course of independent research. This distinction was made by the U.S. 9th Circuit Court of Appeals when evaluating the admissibility of experts

This uses demonstrative evidence, which is evidence created in preparation of trial by attorneys or paralegals.

FORENSIC SCIENCE IN FICTION

Sherlock Holmes, the fictional character created by Sir Arthur Conan Doyle in works produced from 1887 to 1915, used forensic science as one of his investigating methods. Conan Doyle credited the inspiration for Holmes on his teacher at the medical school of the University of Edinburgh, the gifted surgeon and forensic detective Joseph Bell.

Decades later, the comic strip *Dick Tracy* also featured a detective using a considerable number of forensic methods, although sometimes the methods were more fanciful than actually possible.

Defense attorney Perry Mason occasionally used forensic techniques, both in the novels and television series.

Popular television series focusing on crime detection, including *CSI, Cold Case, Bones, Law & Order, NCIS, Criminal Minds, Silent Witness, Dexter,* and *Waking the Dead,* depict glamorized versions of the activities of 21st century forensic scientists. These related TV shows have changed individuals' expectations of forensic science, an influence termed the "CSI effect"

FORENSIC SCIENCE BEING CHALLENGED

Questions about forensic science, fingerprint evidence and the assumption behind these disciplines have been brought to light in some publications, the latest being an article in the New York Post. The article stated boldly that "No one has proved even the basic assumption: That everyone's fingerprint is unique." The article also stated that "Now such assumptions are being questioned — andwith it may come a radical change in how forensic science is used by police departments and prosecutors."

COMPUTER FORENSICS

Computer forensics is a branch of forensic science pertaining to legal evidence found in computers and digital storage mediums. Computer forensics is also known as *digital forensics.*

The goal of computer forensics is to explain the current state of a *digital artifact.* The term digital artifact can include a computer system, a storage medium (such as a hard disk or CD-ROM), an electronic document (e.g. an email message or JPEG image) or even a sequence of packets moving over a computer network. The explanation can be as straightforward as "what information is here?" and as detailed as "what is the sequence of events responsible for the present situation?"

The field of Computer Forensics also has sub branches within it such as Firewall Forensics, Database Forensics and Mobile Device Forensics.

There are many reasons to employ the techniques of computer forensics:

- In legal cases, computer forensic techniques are frequently used to analyze computer systems belonging to defendants (in criminal cases) or litigants (in civil cases).
- To recover data in the event of a hardware or software failure.
- To analyze a computer system after a break-in, for example, to determine how the attacker gained access and what the attacker did.
- To gather evidence against an employee that an organization wishes to terminate.
- To gain information about how computer systems work for the purpose of debugging, performance optimization, or reverse-engineering.

Special measures should be taken when conducting a forensic investigation if it is desired for the results to be used in a court of law. One of the most important measures is to assure that the evidence has been accurately collected and that there is a clear chain of custody from the scene of the crime to the investigator—and ultimately to the court. In order to comply with the need to maintain the integrity of digital evidence, British examiners comply with the Association of Chief Police Officers (A.C.P.O.) guidelines. These are made up of four principles as follows:

- *Principle 1:* No action taken by law enforcement agencies or their agents should change data held on a computer or storage media which may subsequently be relied upon in court.
- *Principle 2:* In exceptional circumstances, where a person finds it necessary to access original data held on a computer or on storage media, that person must be competent to do so and be able to give evidence explaining the relevance and the implications of their actions.
- *Principle 3:* An audit trail or other record of all processes applied to computer based electronic evidence should be created and preserved. An independent third party should be able to examine those processes and achieve the same result.
- *Principle 4:* The person in charge of the investigation (the case officer) has overall responsibility for ensuring that the law and these principles are adhered to.

THE FORENSIC PROCESS

There are five basic steps to the computer forensics:

1. Preparation (of the investigator, not the data)
2. Collection (the data)
3. Examination
4. Analysis
5. Reporting

The investigator must be properly trained to perform the specific kind of investigation that is at hand.

Tools that are used to generate reports for court should be validated. There are many tools to be used in the process. One should determine the proper tool to be used based on the case.

Collecting Digital Evidence

Digital evidence can be collected from many sources. Obvious sources include computers, cell phones, digital cameras, hard drives, CD-ROM, USB memory devices, and so on. Non-obvious sources include settings of digital thermometers, black boxes inside automobiles, RFID tags, and web pages (which must be preserved as they are subject to change).

Special care must be taken when handling computer evidence: most digital information is easily changed, and once changed it is usually impossible to detect that a change has taken place (or to revert the data back to its original state) unless other measures have been taken. For this reason it is common practice to calculate a cryptographic hash of an evidence file and to record that hash elsewhere, usually in an investigator's notebook, so that one can establish at a later point in time that the evidence has not been modified since the hash was calculated.

Other specific practices that have been adopted in the handling of digital evidence include:

- Imaging computer media using a writeblocking tool to ensure that no data is added to the suspect device.
- Establish and maintain the chain of custody.
- Documenting everything that has been done.
- Only use tools and methods that have been tested and evaluated to validate their accuracy and reliability.

Some of the most valuable information obtained in the course of a forensic examination will come from the computer user. An interview with the user can yield valuable information about the system configuration, applications, encryption keys and methodology. Forensic analysis is much easier when analysts have the user's passphrases to access encrypted files, containers, and network servers.

In an investigation in which the owner of the digital evidence has not given consent to have his or her media examined (as in some criminal cases) special care must be taken to ensure that the forensic specialist has the legal authority to seize, copy, and examine the data. Sometimes authority stems from a search warrant. As a general rule, one should not examine digital information unless one has the legal authority to do so. Amateur forensic examiners should keep this in mind before starting any unauthorized investigation.

Live vs. Dead Analysis

Traditionally computer forensic investigations were performed on data at rest—for example, the content of hard drives. This can be thought of as a dead analysis. Investigators were told to shut down computer systems when they were impounded for fear that digital time-bombs might cause data to be erased.

In recent years there has increasingly been an emphasis on performing analysis on live systems. One reason is that many current attacks against computer systems leave no trace on the computer's hard drive—the attacker only exploits information in the computer's memory. Another reason is the growing use of cryptographic storage: it may be that the only copy of the keys to decrypt the storage are in the computer's memory, turning off the computer will cause that information to be lost.

Imaging Electronic Media (evidence)

The process of creating an exact duplicate of the original evidentiary media is often called Imaging. Using a standalone hard-drive duplicator or software imaging tools such as DCFLdd or IXimager, the *entire* hard drive is completely duplicated. This is usually done at the sector level, making a bit-stream copy of every part of the user-accessible areas of the hard drive which can physically store data, rather than duplicating the filesystem. The original drive is then moved to secure storage to prevent tampering. During imaging, a write protection device or application is normally used to ensure that no information is introduced onto the evidentiary media during the forensic process.

The imaging process is verified by using the SHA-1 message digest algorithm (with a program such as sha1sum) or other still viable algorithms such as MD5. At critical points throughout the analysis, the media is verified again, known as "hashing", to ensure that the evidence is still in its original state. In corporate environments seeking civil or internal charges, such steps are generally overlooked due to the time required to perform them. They are essential for evidence that is to be presented in a court room, however.

Collecting Volatile Data

If the machine is still active, any intelligence which can be gained by examining the applications currently open is recorded. If the machine is suspected of being used for illegal communications, such as terrorist traffic, not all of this information may be stored on the hard drive. If information stored solely in RAM is not recovered before powering down it may be lost. This results in the need to collect volatile data from the computer at the onset of the response.

Several Open Source tools are available to conduct an analysis of open ports, mapped drives (including through an active VPN connection), and open or mounted encrypted files (containers) on the live computer system. Utilizing open source tools and commercially available products, it is possible to obtain an image of these mapped drives and the open encrypted containers in an unencrypted format. Open Source tools for PCs include Knoppix and Helix. Commercial imaging tools include Access Data's Forensic Toolkit and Guidance Software's EnCase application.

The aforementioned Open Source tools can also scan RAM and Registry information to show recently accessed web-based email sites and the login/password combination used. Additionally these tools can also yield login/password for recently accessed local email applications including MS Outlook.

In the event that partitions with EFS are suspected to exist, the encryption keys to access the data can also be gathered during the collection process. With Microsoft's most recent addition, Vista, and Vista's use of BitLocker and the Trusted Platform Module (TPM), it has become necessary in some instances to image the logical hard drive volumes before the computer is shut down.

RAM can be analyzed for prior content after power loss. Although as production methods become cleaner the impurities used to indicate a particular cell's charge prior to power loss are becoming less common. However, data held statically in an area of RAM for long periods of time are more likely to be detectable using these methods. The likelihood of such recovery increases as the originally applied voltages, operating temperatures and duration of data storage increases. Holding unpowered RAM below - 60 □C will help preserve the residual data by an order of magnitude, thus improving the chances of successful recovery. However, it can be impractical to do this during a field examination.

Analysis

All digital evidence must be analyzed to determine the type of inform n many investigations, numerous other tools are used to analyze specific portions of information.

Reporting

Once the analysis is complete, a report is generated. This report may be a written report, oral testimony, or some combination of the two.

COMPARISON TO PHYSICAL FORENSICS

There are many core differences between computer forensics and "physical forensics." At the highest level, the physical forensic science e or blood?) or the source of the item (i.e. did this blood come from person X?). Computer forensics on the other hand focuses on finding the evidence and analyzing it. Therefore, it is more analogous to a physical crime scene investigation than the physical forensic processes.

EXAMPLES OF COMPUTER FORENSICS

Computer forensics has played a pivotal role in many cases.

Chandra Levy

Chandra Levy was a Washington, D.C. intern who disappeared on April 30, 2001. She had used the web and e-mail to make travel arrangements and communicate with her parents. Information found on her computer led police to search most of Rock Creek Park, where her body was eventually found one year later by a man walking his dog.

BTK Killer

Dennis Rader was convicted of a string of serial killings that occurred over a period of sixteen years. Towards the end of this period, Rader sent letters to the police on a floppy disk. Metadata within the documents implicated an author named "Dennis" at "Christ Lutheran Church"; this evidence helped lead to Rader's arrest.

Aside from uncovering evidence to assist the prosecution in high-profile murder cases, an in-depth knowledge of computer forensics is integral when presenting digital evidence for legal proceedings.

CHAPTER - 8

Gravimetric Analysis

Gravimetric analysis describes a set of methods in analytical chemistry for the quantitative determination of an analyte based on the mass of a solid. A simple example is the measurement of solids suspended in a water sample: A known volume of water is filtered, and the collected solids are weighed. In most cases, the analyte must first be converted to a solid by precipitation with an appropriate reagent. The precipitate can then be collected by filtration, washed, dried to remove traces of moisture from the solution, and weighed.

The amount of analyte in the original sample can then be calculated from the mass of the precipitate and its chemical composition. In other cases, it may be easier to remove the analyte by vaporization. The analyte might be collected — perhaps in a cryogenic trap or on some absorbent material such as activated carbon—and measured directly. Or, the sample can be weighed before and after it is dried; the difference between the two masses gives the mass of analyte lost. This is especially useful in determining the water content of complex materials such as foodstuffs.

Fig. 8.1: Analytical Balance

PROCEDURE

1. The sample is dissolved, if it is not already in solution.
2. The solution may be treated to adjust the pH (so that the proper precipitate is formed, or to suppress the formation of other precipitates). If it is known that species are present which interfere (by also forming precipitates under the same conditions as the analyte), the sample might require treatment with a different reagent to remove these interferents.
3. The precipitating reagent is added at a concentration that favours the formation of a "good" precipitate. This

may require low concentration, extensive heating (often described as "digestion"), or careful control of the pH. Digestion can help reduce the amount of coprecipitation.

4. After the precipitate has formed and been allowed to "digest", the solution is carefully filtered. The filter is chosen to trap the precipitate; smaller particles are more difficult to filter.

   - Depending on the procedure followed, the filter might be a piece of ashless filter paper in a fluted funnel, or a filter crucible. Filter paper is convenient because it does not typically require cleaning before use; however, filter paper can be chemically attacked by some solutions (such as concentrated acid or base), and may tear during the filtration of large volumes of solution.
   - The alternative is a crucible whose bottom is made of some porous material, such as sintered glass, porcelain or sometimes metal. These are chemically inert and mechanically stable, even at elevated temperatures. However, they must be carefully cleaned to minimize contamination or carryover (cross-contamination). Crucibles are often used with a mat of glass or asbestos fibers to trap small particles.
   - After the solution has been filtered, it should be tested to make sure that the analyte has been completely precipitated. This is easily done by adding a few drops of the precipitating reagent; if a precipitate is observed, the precipitation is incomplete.

5. After filtration, the precipitate — including the filter paper or crucible — is heated. This achieves three purposes:

   - The remaining moisture is removed (drying).

- Secondly, the precipitate is converted to a more chemically stable form. For instance, calcium ion might be precipitated using oxalate ion, to produce calcium oxalate (CaC_2O_4); it might then be heated to convert it into the oxide (CaO). It is vital that the empirical formula of the weighed precipitate be known, and that the precipitate be pure; if two forms are present, the results will be inaccurate.
- The precipitate cannot be weighed with the necessary accuracy in place on the filter paper; nor can the precipitate be completely removed from the filter paper in order to weigh it. The precipitate can be carefully heated in a crucible until the filter paper has burned away; this leaves only the precipitate. (As the name suggests, "ashless" paper is used so that the precipitate is not contaminated with ash.)

6. After the precipitate is allowed to cool (preferably in a desiccator to keep it from absorbing moisture), it is weighed (in the crucible). The mass of the crucible is subtracted from the combined mass, giving the mass of the precipitated analyte. Since the composition of the precipitate is known, it is simple to calculate the mass of analyte in the original sample.

Washing and Filtering

The precipitate is often washed to remove impurities adsorbed onto the surface of the particles. Washing may be done with a solution of the precipitating agent, to avoid redissolving a slightly soluble salt. With many precipitates, peptization occurs during washing. Here part of the precipitate reverts to the colloidal form (e.g. $AgCl_{(colloidal)} \leftrightarrow AgCl(s)$.) This results in the loss of part of the precipitate because the colloidal form may pass through on filtration. Peptization can be reduced with careful technique and washing with a solution of appropriate pH and ionic strength.

EXAMPLE

A chunk of ore is treated with concentrated nitric acid and potassium chlorate to convert all of the sulphur to sulphate (SO_4^{2-}). The nitrate and chlorate are removed by treating the solution with concentrated HCl. The sulfate is precipitated with barium (Ba^{2+}) and weighed as $BaSO_4$.

ADVANTAGES

Gravimetric analysis, if methods are followed carefully, provides for exceedingly precise analysis. In fact, gravimetric analysis was used to determine the atomic masses of many elements to six figure accuracy. Gravimetry provides very little room for instrumental error and does not require a series of standards for calculation of an unknown. Also, methods often do not require expensive equipment. Gravimetric analysis, due to its high degree of accuracy, when performed correctly, can also be used to calibrate other instruments in lieu of reference standards.

DISADVANTAGES

Gravimetric analysis usually only provides for the analysis of a single element, or a limited group of elements, at a time. Comparing modern dynamic flash combustion coupled with gas chromatography with traditional combustion analysis will show that the former is both faster and allows for simultaneous determination of multiple elements while traditional determination allowed only for the determination of carbon and hydrogen. Methods are often convoluted and a slight mis-step in a procedure can often mean disaster for the analysis (colloid formation in precipitation gravimetry, for example). Compare this with hardy methods such as spectrophotometry and one will find that analysis by these methods is much more efficient.

GRAVIMETRY

Introduction

Gravimetry is the measurement of a gravitational field. Gravimetry may be used when either the magnitude of gravitational field or the properties of matter responsible for its creation are of interest. The term gravimetry or gravimetric is also used in chemistry to define a class of analytical procedures, called gravimetric analysis relying upon weighing a sample of material.

HISTORY

The modern gravimeter was developed by Lucien LaCoste and Arnold Romberg in 1936.

They also invented most subsequent refinements, including the ship-mounted gravimeter, in 1965, temperature-resistant instruments for deep boreholes, and lightweight hand-carried instruments. Most of their designs remain in use (2005) with refinements in data collection and data processing.

UNITS OF MEASUREMENT

Gravity is usually measured in units of acceleration. In the SI system of units, the standard unit of acceleration is 1 metre per second squared (abbreviated as m/s^2). Other units include the gal (sometimes known as a *galileo,* in either case with symbol Gal), which equals 1 centimetre per second squared, and the *g* (*g*n), equal to 9.80665 m/s^2. The value of the *g*n approximately equals the acceleration due to gravity at the Earth's surface (although the actual acceleration *g* varies fractionally from place to place).

How gravity is measured ?

An instrument used to measure gravity is known as a gravimeter, or gravitometer. Since general relativity regards the effects of gravity as indistinguishable from the effects of acceleration, gravimeters may be regarded as special purpose accelerometers. Many weighing scales may be regarded as simple gravimeters. In one common form, a spring is used to counteract the force of gravity pulling on an object. The change in length of the spring may be calibrated to the force required to balance the gravitational pull. The resulting measurement may be made in units of force (such as the newton), but is more commonly made in units of gals.

More sophisticated gravimeters are used when precise measurements are needed. When measuring the Earth's gravitational field, measurements are made to the precision of microgals to find density variations in the rocks making up the Earth. Several types of gravimeters exist for making these measurements, including some that are essentially refined versions of the spring scale described above. These measurements are used to define gravity anomalies.

Besides precision, also stability is an important property of a gravimeter, as it allows the monitoring of gravity *changes*. These changes can be the result of mass displacements inside the Earth, or of vertical movements of the Earth's crust on which measurements are being made: remember that gravity decreases 0.3 mGal for every metre of height. The study of gravity changes belongs to geodynamics.

The majority of modern gravimeters use specially-designed quartz zero-length springs to support the test mass. Zero length springs do not follow Hooke's Law, instead they have a force proportional to their length. The special property of these springs is that a vertical pendulum can be designed with a period approaching a thousand seconds. This detunes the test mass from most local vibration and mechanical noise, increasing the sensitivity and utility of the gravimeter. The springs are quartz so that magnetic and electric fields do not affect measurements. The test mass is sealed in an air-tight container so that tiny changes of barometric pressure from blowing wind and other weather do not change the buoyancy of the test mass in air.

Spring gravimeters are, in practice, relative instruments which measure the difference in gravity between different locations. A relative instrument also requires calibration by comparing instrument readings taken at locations with known complete or absolute values of gravity. Absolute gravimeters provide such measurements by determining the gravitational acceleration of a test mass in vacuum. A test mass is allowed to fall freely inside a vacuum chamber and its position is measured with a laser interferometer and timed with an atomic clock. The laser wavelength is known to ±0.025 ppb and the clock is stable to ±0.03 ppb as well. Great care must be taken to minimize the effects of perturbing forces such as residual air resistance (even in vacuum) and magnetic forces. Such instruments are capable of an accuracy of a few parts per billion or 0.002 mGal and reference their measurement to atomic standards of length and time. Their primary use is for calibrating relative instruments, monitoring crustal deformation, and in geophysical studies requiring high accuracy and stability. However, absolute instruments are somewhat larger and significantly more expensive than relative spring gravimeters, and are thus relatively rare.

Gravimeters have been designed to mount in vehicles, including aircraft, ships and submarines. These special gravimeters isolate acceleration from the movement of the vehicle, and subtract it from measurements. The acceleration of the vehicles is often hundreds or thousands of times stronger than the changes being measured. A gravimeter (the *Lunar Surface Gravimeter*) was also deployed on the surface of the moon during the Apollo 17 mission, but did not work due to a design error. A second device (the *Traverse Gravimeter Experiment*) functioned as anticipated.

MICROGRAVIMETRY

Microgravimetry is a rising and important branch developed on the foundation of classical gravimetry. Microgravity investigations are carried out in order to solve various problems of engineering geology, mainly location of voids and their monitoring. Very detailed measurements of high accuracy can indicate voids of any origin, provided the size and depth are large enough to produce gravity effect stronger than is the level of confidence of relevant gravity signal.

CHAPTER - 9

Electroanalytical Method

INTRODUCTION

Electroanalytical methods are a class of techniques in analytical chemistry which study an analyte by measuring the potential (Volts) and/or current (Amps) in an electrochemical cell containing the analyte. These methods can be broken down into several categories depending on which aspects of the cell are controlled and which are measured. The three main categories are Potentiometry (the difference in electrode potentials is measured), Coulometry (the cell's current is measured over time), and Voltammetry (the cell's current is measured while actively altering the cell's potential).

POTENTIOMETRY

Potentiometry passively measures the potential of a solution between two electrodes, affecting the solution very little in the process. The potential is then related to the concentration of one or more analytes. The cell structure used is often referred to as an electrode even though it actually contains *two* electrodes: an *indicator electrode* and a *reference electrode* (distinct from the reference electrode used in the three electrode system). Potentiometry usually uses electrodes made *selectively* sensitive to the ion of interest, such as a fluoride-selective electrode. The most common potentiometric electrode is the glass-membrane electrode used in a pH meter.

COULOMETRY

Coulometry uses applied current or potential to completely convert an analyte from one oxidation state to another. In these experiments, the total current passed is measured directly or indirectly to determine the number of electrons passed. Knowing the number of electrons passed can indicate the concentration of the analyte or, when the concentration is known, the number of electrons transferred in the redox reaction. Common forms of coulometry include bulk electrolysis, also known as *Potentiostatic coulometry* or *controlled potential coulometry,* as well as a variety of coulometric titrations.

VOLTAMMETRY

Voltammetry applies a constant and/or varying potential at an electrode's surface and measures the resulting current with a three electrode system. This method can reveal the reduction potential of an analyte and its electrochemical reactivity. This method in practical terms is nondestructive since only a very small amount of the analyte is consumed at the two-dimensional surface of the working and auxiliary electrodes. In practice the analyte solutions is usually disposed of since it is difficult to separate the analyte from the bulk electrolyte and the experiment requires a small amount of analyte. A normal experiment may involve 1-10 mL solution with an analyte concentration between 1-10 mM.

POLAROMETRY

Polarometry is a subclass of voltammetry that uses a dropping mercury electrode as the working electrode. The auxiliary electrode is often the resulting mercury pool. Concern over the toxicity of mercury has caused the use of mercury electrodes to decrease greatly. Alternate electrode materials, such as the noble metals and glassy carbon, are affordable, inert, and easily cleaned.

AMPEROMETRY

Most of Amperometry is now a subclass of voltammetry in which the electrode is held at constant potentials for various lengths of time. The distinction between *amperometry* and *voltammetry* is mostly historic. There was a time when it was difficult to switch between "holding" and "scanning" a potential. This function is trivial for modern potentiostats, and today there is little distinction between the techniques which either "hold", "scan", or do both during a single experiment. Yet the terminology still results in confusion, for example, differential pulse voltammetry is also referred to as and *differential pulse amperometry*. This experiment can be seen as the combination of linear sweep voltammetry and chronoam-perometry thus the confusion in which category it should be named.

One advantage that distinguishes amperometry from other forms of voltammetry is that in amperometry, the currents are averaged (or summed) over time. In most of voltammetry, current readings must be considered independently at individual time intervals. The averaging used in amperometry gives these methods greater precision than the many individual readings of (other) voltammetric techniques.

Not all of the experiments which were historically amperometry now fall under the domain of voltammetry. In an amperometric titration, the current is measured, but this would not be considered voltammetry since the entire solution is transformed during the experiment. Amperometric titrations are instead a form of coulometry.

ION SELECTIVE ELECTRODE

An Ion-selective electrode (ISE) (also known as a specific ion electrode (SIE)) is a transducer (sensor) which converts the activity of a specific ion dissolved in a solution into an electrical potential which can be measured by a voltmeter or pH meter. The voltage is theoretically dependent on the logarithm of the ionic activity, according to the Nernst equation. The sensing part of the electrode is usually made as an ion-specific membrane, along with a reference electrode. Ion-selective electrodes are used in biochemical and biophysical research, where measurements of ionic concentration in an aqueous solution are required, usually on a real time basis.

TYPES OF ION-SELECTIVE MEMBRANE

There are four main types of ion-selective membrane used in ion-selective electrodes:

1. Glass;
2. Solid state;
3. Liquid based; and
4. Compound electrode.

Glass Membranes

Glass membranes are made from an ion-exchange type of glass (silicate of chalcogenide). This type of ISE has good selectivity, but only for several single-charged cations; mainly H^+, Na^+, and Ag^+. Chalcogenide glass also has selectivity for double-charged metal ions, such as Pb^{2+}, and Cd^{2+}. The glass membrane has excellent chemical durability and can work in very aggressive media. A very common example of this type of electrode is the pH glass electrode.

Crystalline Membranes

Crystalline membranes are made from mono- or polycrystallites of a single substance. They have good selectivity, because only ions which can introduce themselves into the crystal structure can interfere with the electrode response. Selectivity of crystalline membranes can be for both cation and anion of the membrane-forming substance. An example is the fluoride selective electrode based on LaF_3 crystals.

Ion Exchange Resin Membranes

Ion-exchange resins are based on special organic polymer membranes which contain a specific ion-exchange substance (resin). This is the most widespread type of ion-specific electrode. Usage of specific resins allows preparation of selective electrodes for tens of different ions, both single-atom or multi-atom. They are also the most widespread electrodes with anionic selectivity. However, such electrodes have low chemical and physical durability as well as "survival time". An example is the potassium selective electrode, based on valinomycin as an ion-exchange agent.

Construction

These electrodes are prepared from glass capillary tubing approximately 2 millimeters in diameter, a large batch at a time. Polyvinyl chloride is dissolved in a solvent and plasticizers (typically phthalates) added, in the standard fashion used when making something out of vinyl. In order to provide the ionic specificity, a specific ion channel or carrier is added to the solution; this allows the ion to pass through the vinyl, which prevents the passage of other ions and water.

One end of a piece of capillary tubing about an inch or two long is dipped into this solution and removed to let the vinyl solidify into a plug at that end of the tube. Using a syringe and needle, the tube is filled with salt solution from the other end, and may be stored in a bath of the salt solution for an indeterminate period. For convenience in use, the open end of the tubing is fitted through a tight o-ring into a somewhat larger diameter tubing containing the same salt solution, with a silver or platinum electrode wire inserted. New electrode tips can thus be changed very quickly by simply removing the older electrode and replacing it with a new one.

APPLICATION

In use, the electrode wire is connected to one terminal of a galvanometer or pH meter, the other terminal of which is connected to a reference electrode, and both electrodes are immersed in the solution to be tested. The passage of the ion through the vinyl via the carrier or channel creates an electrical current, which registers on the galvanometer; by calibrating against standard solutions of varying concentration, the ionic concentration in the tested solution can be estimated from the galvanometer reading.

In practice there are several issues which affect this measurement, and different electrodes from the same batch will differ in their properties. Leakage between the vinyl and the wall of the capillary, thereby allowing passage of any ions, will cause the meter reading to show little or no change between the various calibration solutions, and requires that that electrode be discarded. Similarly, with use the ion-sensitive channels in the vinyl appear to gradually become blocked or otherwise inactivated, causing the electrode to lose sensitivity. The response of the electrode and galvanometer is temperature sensitive, and also 'drifts' over time, requiring recalibration frequently during a series of measurements, ideally at least one calibration sample before and after each test sample. On the other hand, after immersion in the solution there is a transient 'settling time' which can be five minutes or even longer, before the electrode and galvanometer equilibrate to a new reading; so that timing of the reading is critical in order to find the most accurate 'window' after the response has settled, but before it has drifted appreciably.

ENZYME ELECTRODES

Enzyme electrodes definitely are not true ion-selective electrodes but usually are considered within the ion-specific electrode topic. Such an electrode has a "double reaction" mechanism - an enzyme reacts with a specific substance, and the product of this reaction (usually H^+ or OH^-) is detected by a true ion-selective electrode, such as a pH-selective electrodes. All these reactions occur inside a special membrane which covers the true ion-selective electrode, which is why enzyme electrodes sometimes are considered as ion-selective. An example is glucose selective electrodes.

INTERFERENCE

The most serious problem limiting use of ion-selective electrodes is interference from other, undesired, ions. No ion-selective electrodes are completely ion-specific; all are sensitive to other ions having similar physical properties, to an extent which depends on the degree of similarity. Most of these interferences are weak enough to be ignored, but in some cases the electrode may actually be much more sensitive to the interfering ion than to the desired ion, requiring that the interfering ion be present only in relatively very low concentrations, or entirely absent. In practice, the relative sensitivities of each type of ion-specific electrode to various interfering ions is generally known and should be checked for each case; however the precise degree of interference depends on many factors, preventing precise correction of readings. Instead, the calculation of relative degree of interference from the concentration of interfering ions can only be used as a guide to determine whether the approximate extent of the interference will allow reliable measurements, or whether the experiment will need to be redesigned so as to reduce the effect of interfering ions.

CHAPTER - 10

Coulometry

INTRODUCTION

Coulometry is the name given to a group of techniques in analytical chemistry that determine the amount of matter transformed during an electrolysis reaction by measuring the amount of electricity (in coulombs) consumed or produced.

There are two basic categories of coulometric techniques. Potentiostatic coulometry involves holding the electric potential constant during the reaction using a potentiostat. The other, called coulometric titration or amperostatic coulometry, keeps the current (measured in amperes) constant using an amperostat.

POTENTIOSTATIC COULOMETRY

Potentiostatic coulometry is a technique most commonly referred to as "bulk electrolysis". The working electrode is kept at a constant potential and the current which flows through the circuit is measured. This constant potential is applied long enough to fully reduce or oxidize all of the substrate in a given solution. As the substrate is consumed, the current also decreases, approaching zero when the conversion is complete. The sample mass, molecular mass, number of electrons in the electrode reaction, and number of electrons passed during the experiment are all related by Faraday's laws. It follows that, if three of the values are known, then the fourth can be calculated.

Bulk electrolysis is often used to unambiguously assign the number of electrons consumed in a reaction observed through voltammetry. It also has the added benefit of producing a solution of a species (oxidation state) which may not be accessible through chemical routes. This species can then be isolated or further characterized while in solution.

The rate of such reactions is not determined by the concentration of the solution, but rather the mass transfer of the substrate in the solution to the electrode surface. Rates will increase when the volume of the solution is decreased, the solution is stirred more rapidly, or the area of the working electrode is increased. Since mass transfer is so important the solution is stirred during a bulk electrolysis. However, this technique is generally not considered a hydrodynamic technique, since a laminar flow of solution against the electrode is neither the objective nor outcome of the stirring.

The extent to which a reaction goes to completion is also related to how much greater the applied potential is than the reduction potential of interest. In the case where multiple reduction potentials are of interest, it is often difficult to set an electrolysis potential a "safe" distance (such as 200 mV) past a redox event. The result is incomplete conversion of the substrate, or else conversion of some of the substrate to the more reduced form. This factor must be considered when analyzing the current passed and when attempting to do further analysis/isolation/experiments with the substrate solution.

An advantage to this kind of analysis over electrogravimetry is that it does not require that the product of the reaction be weighed. This is useful for reactions where the product does not deposit as a solid, such as the determination of the amount of arsenic in a sample from the electrolysis of arsenous acid (H_3AsO_3) to arsenic acid (H_3AsO_4).

COULOMETRIC TITRATION

Coulometric titrations use a constant current system to accurately quantify the concentration of a species. In this experiment, the applied current is equivalent to a titrant. Current is applied to the unknown solution until all of the unknown species is either oxidized or reduced to a new state, at which point the potential of the working electrode shifts dramatically. This potential shift indicates the end point. The magnitude of the current (in amps) and the duration of the current (seconds) can be used to determine the moles of the unknown species in solution. When the volume of the solution is known, then the molarity of the unknown species can be determined.

APPLICATIONS

Karl Fischer Reaction

The Karl Fischer reaction uses a coulometric titration to determine the amount of water in a sample. It can determine concentrations of water on the order of mg/L. It is used to find the amount of water in substances such as butter, sugar, cheese, paper, and petroleum.

The reaction involves converting solid iodine into hydrogen iodide in the presence of sulfur dioxide and water. Methanol is most often used as the solvent, but ethylene glycol and diethylene glycol also work. Pyridine is often used to prevent the build up of sulfuric acid, although the use of imidazole and diethanolamine for this role are becoming more common. All reagents must be anhydrous for the analysis to be quantitative. The balanced chemical equation, using methanol and pyridine, is:

$$[C_5H_5NH]SO_3CH_3^- + I_2 + H_2O + 2C_5H_5N \longrightarrow$$
$$[C_5H_5NH]SO_4CH_3 + 2\,[C_5H_5NH]I$$

In this reaction, a single molecule of water reacts with a molecule of iodine. Since this technique is used to determine the water content of samples, atmospheric humidity could alter the results. Therefore, the system is usually isolated with drying tubes or placed in an inert gas container. In addition, the solvent will undoubtedly have some water in it so the solvent's water content must be measured to compensate for this inaccuracy.

To determine the amount of water in the sample, analysis must first be performed using either back or direct titration. In the direct method, just enough of the reagents will be added to completely use up all of the water. At this point in the titration, the current approaches zero. It is then possible to relate the amount of reagents used to the amount of water in the system via stoichiometry. The back-titration method is similar, but involves the addition of an excess of the reagent. This excess is then consumed by adding a known amount of a standard solution with known water content. The result reflects the water content of the sample and the standard solution. Since the amount of water in the standard solution is known, the difference reflects the water content of the sample.

Determination of Film Thickness

Coulometry can be used in the determination of the thickness of metallic coatings. This is performed by measuring the quantity of electricity needed to dissolve a well-defined area of the coating. The film thickness is proportional to the constant current i, the molecular weight M of the metal, the density of the metal, and the surface area A:

$$\Delta \alpha \frac{iM}{Ap}$$

The electrodes for this reaction are often platinum electrode and an electrode that relates to the reaction. For tin coating on a copper wire, a tin electrode is used, while a sodium chloride-zinc sulfate electrode would be used to determine the zinc film on a piece of steel. Special cells have been created to adhere to the surface of the metal to measure its thickness. These are basically columns with the internal electrodes with magnets or weights to attach to the surface. The results obtained by this coulometric method are similar to those achieved by other chemical and metallurgic techniques.

COULOMETERS

Electronic Coulometer

The electronic coulometer is based on the application of the Operational amplifier in the "integrator"-type circuit. The current passed through the resistor R1 makes a potential drop which is integrated by operational amplifier on the capacitor plates; the higher current, the larger the potential drop. The current need not be constant. In such scheme V_{out} is proportional of the passed charge (i*t). Sensitivity of the coulometer can be changed by choosing of the appropriate value of R_1.

Electrochemical Coulometers

There are three common types of coulometers based on electrochemical processes:

- Copper coulometer
- Mercury coulometer
- Hofmann voltameter

"Voltameter" is a synonym for "coulometer".

HOFMANN VOLTAMETER

Introduction

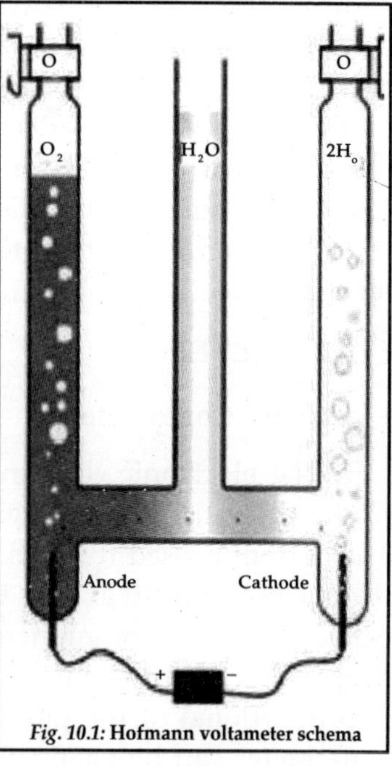

Fig. 10.1: Hofmann voltameter schema

A Hofmann voltameter is an apparatus for electrolyzing water, invented by August Wilhelm von Hofmann (1818–1892) It consists of three joined upright cylinders, usually glass. The inner cylinder is open at the top to allow addition of water and an ionic compound to improve conductivity, such as a small amount of sulphuric acid. A platinum electrode is placed inside the bottom of each of the two side cylinders, connected to the positive and negative terminals of a source of electricity. When current is run through Hofmann's Voltameter, gaseous oxygen forms at the anode and gaseous hydrogen at the cathode. Each gas displaces water and collects at the top of the two outer tubes.

MATCH TEST

The presence of gas in the collecting tube can be detected with a smoldering match or "glowing wood splint"; oxygen will cause the match to immediately burst into a bright white flame and will burn vigorously, while the presence of hydrogen gas will create a "pop" sound (a very small explosion) due to the hydrogen rapidly burning.

NAME

The name 'voltameter' was coined by Daniell who shortened Faraday's original name of "volta-electrometer".

Hofmann voltameters are no longer used as electrical measuring devices. However, before the invention of the ammeter, voltameters were often used to measure direct current, since current through a voltameter with iron or copper electrodes electroplates the cathode with an amount of metal from the anode directly proportional to the total current (Faraday's law of electrolysis). The modern name is "electrochemical coulometer".SHU ABAS

USES

The amount of electricity that has passed through the system can then be determined by weighing the cathode. Thomas Edison used voltameters as electricity meters. (A Hofmann voltameter cannot be used to weigh electric current in this fashion, as the platinum electrodes are too inert for plating.)

A Hofmann voltameter is often used as a demonstration of stoichiometric principles, as the two-to-one ratio of the volumes of hydrogen and oxygen gas produced by the apparatus illustrates the chemical formula of water, H_2O.

CHAPTER - 11

Voltammetry

Voltammetry is a category of electroanalytical methods used in analytical chemistry and various industrial processes. In voltammetry, information about an analyte is obtained by measuring the current as the potential is varied.

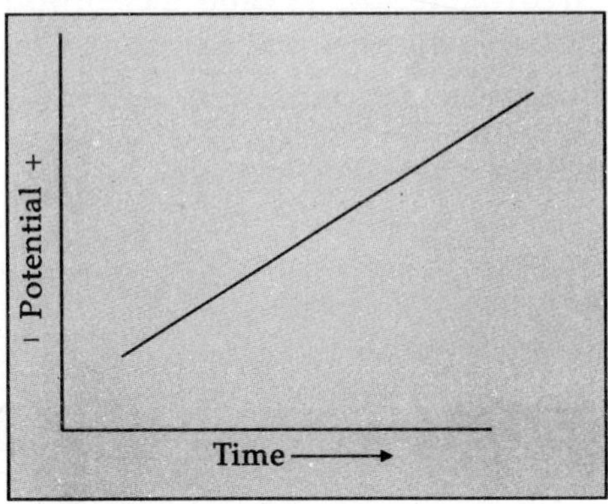

Fig 11.1: Linear potential sweep

THREE ELECTRODE SYSTEM

Voltammetry experiments investigate the half cell reactivity of an analyte. Most experiments control the potential (volts) of an electrode in contact with the analyte while measuring the resulting current (amperes).

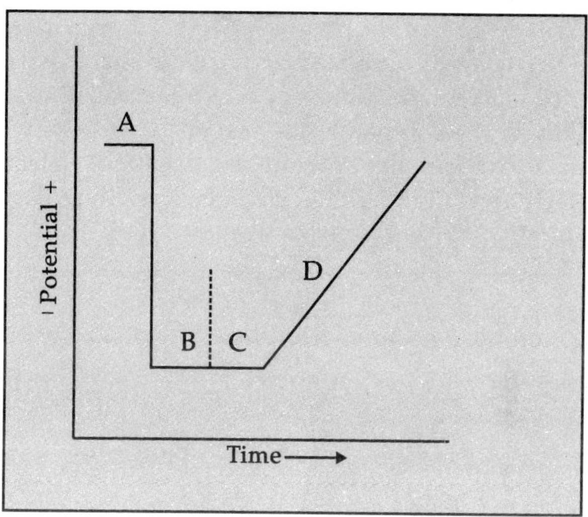

Fig 11.2: Potential as a function of time for anodic stripping voltammetry

To conduct such an experiment requires at least two electrodes. The working electrode, which makes contact with the analyte, must apply the desired potential in a controlled way and facilitate the transfer of electrons to and from the analyte. A second electrode acts as the other half of the cell. This second electrode must have a known potential with which to gauge the potential of the working electrode, furthermore it must balance the electrons added or removed by the working electrode. While this is a viable setup, it has a number of shortcomings. Most significantly, it is extremely difficult for an electrode to maintain a constant potential while passing current to counter redox events at the working electrode.

To solve this problem, the role of supplying electrons and referencing potential has been divided between two separate electrodes. The reference electrode is a half cell with a known reduction potential. Its only role is to act as reference in measuring and controlling the working electrodes potential and at no point does it pass any current. The auxiliary electrode passes all the current needed to balance the current observed at the working electrode. To achieve this current, the auxiliary will often swing to extreme potentials at the edges of the solvent window, where it oxidizes or reduces the solvent or supporting electrolyte. These electrodes, the working, reference, and auxiliary make up the modern three electrode system.

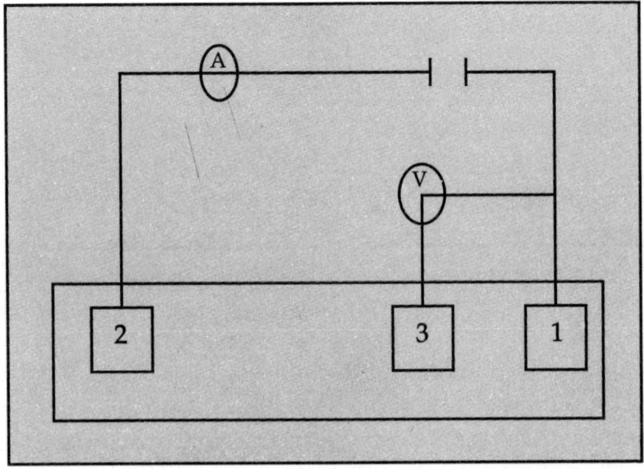

Fig 11.3: There electrode set-up: (1) working electrode; (2) auxiliary electrode; (3) reference electrode

There are many systems which have more electrodes, but their design principles are generally the same as the three electrode system. For example, the rotating ring-disk electrode has two distinct and separate working electrodes, a disk and a ring, which can be used to scan or hold potentials independently of each other. Both of these electrodes are balanced by a single reference and auxiliary combination for an over all four electrode design. More complicated experiments may add working electrodes as required and at times reference or auxiliary electrodes.

In practice it can be very important to have a working electrode with known dimensions and surface characteristics.

As a result, it is common to clean and polish working electrodes regularly. The auxiliary electrode can be almost anything as long as it doesn't react with the bulk of the anaylte solution and conducts well. The reference is the most complex of the three electrodes, there are a variety of standards used and its worth investigating elsewhere. For non-aqueous work, IUPAC recommends the use of the ferrocene/ferrocenium couple as an internal standard. In most voltammetry experiments a bulk electrolyte (also known as supporting electrolyte) is used to minimize solution resistance. It can be possible to run an experiment without an bulk electrolyte but the added resistance greatly reduces accuracy of the results. In the case of room temperature ionic liquids the solvent can act as the electrolyte.

THEORY

The Nernst equation is fundamental to voltammetry and can be used for a reversible reaction. In this equation the R represents the reduced species and O the oxidized.

$$E = E^0 - \frac{RT}{nF} \ln \frac{c_R^0}{c_O^0}$$

R = molar gas constant

T = absolute temperature

n = number of electrons transferred (as determined by the stoichiometry of the cell reaction)

F = Faraday constant

E = applied potential

E^0 = standard reduction potential

Another useful equation in voltammetry is the Butler-Volmer equation. This equation represents the relationship between concentration, current, and potential.

$$\frac{i}{nFA} = k^0 \left\{ c_O^0 \exp[-\alpha\theta] - c_R^0 \exp[(1-\alpha)\theta] \right\}$$

$= nF(E - E0)/RT$

k^0 = heterogeneous rate constant

a = transfer coefficient

A = area of the electrode

i = current

n = number of electrons transferred

TYPES OF VOLTAMMETRY

- *Linear sweep voltammetry*
- *Staircase voltammetry*
- *Squarewave voltammetry*
- *Cyclic voltammetry* - A voltammetric method that can be used to determine diffusion coefficients and half cell reduction potentials.
- *Anodic stripping voltammetry* - A quantitative, analytical method for trace analysis of metal cations. The analyte is deposited (electroplated) onto the working electrode during a deposition step, and then oxidized during the stripping step. The current is measured during the stripping step.
- *Cathodic stripping voltammetry* - A quantitative, analytical method for trace analysis of anions. A positive potential is applied, oxidizing the mercury electrode and forming insoluble precipitates of the anions. A negative potential then reduces (strips) the deposited film into solution.
- *Adsorptive stripping voltammetry* - A quantitative, analytical method for trace analysis. The analyte is deposited simply by adsorption on the electrode surface (i.e., no electrolysis), then electrolyzed to give the analytical signal. Chemically modified electrodes are often used.
- *Alternating current voltammetry.*

- *Polarography* - a subclass of voltammetry where the working electrode is a dropping mercury electrode (DME), useful for its wide cathodic range and renewable surface.
- *Rotated electrode voltammetry* - A hydrodynamic technique in which the working electrode, usually a rotating disk electrode (RDE) or rotating ring-disk electrode (RRDE), is rotated at a very high rate. This technique is useful for studying the kinetics and electrochemical reaction mechanism for a half reaction.
- *Normal pulse voltammetry.*
- *Differential pulse voltammetry.*
- *Chronoamperometry.*

HISTORY

The beginning of voltammetry was facilitated by the discovery of polarography in 1922 by the Nobel Prize winning chemist Jaroslav Heyrovský. Early voltammetric techniques had many problems, limiting their viability for everyday use in analytical chemistry. The 1960s and 1970s saw many advances in the theory and instrumentation of all voltammetric methods. These advancements improved sensitivity and created new analytical methods. Industry responded with the production of cheaper instruments that could be effectively used in routine analytical work.

APPLICATIONS

Voltametric sensors A number of voltammetric systems are produced commercially for the determination of specific species that are of interest in industry and research. These devices are sometimes called electrodes but are, in fact, complete voltammetric cells and are better referred to as sensors.

The oxygen electrode The determination of dissolved oxygen in a variety of aqueous environments, such as sea water, blood, sewage, effluents from chemical plants, and soils is of tremendous importance to industry, biomedical and environmental research, and clinical medicine. One of the most common and convenient methods for making such measurements is with the Clark oxygen sensor, which was patented by L.C. Clark, Jr. in 1956.

POLAROGRAPHY

Polarography is a subclass of voltammetry where the working electrode is a dropping mercury electrode (DME), useful for its wide cathodic range and renewable surface. It was invented by Jaroslav Heyrovský.

THEORY OF OPERATION

Polarography is an voltammetric measurement whose response is determined by combined diffusion/convection mass transport. Polarography is a specific type of measurement that falls into the general category of linear-sweep voltammetry where the electrode potential is altered in a linear fashion from the initial potential to the final potential. As a linear sweep method controlled by convection/diffusion mass transport, the current vs. potential response of a polarographic experiment has the typical sigmoidal shape. What makes polarography different from other linear sweep voltammetry measurements is that polarography makes use of the dropping mercury electrode (DME). Close-up of the DME.

A plot of the current vs. potential in a polarography experiment shows the current oscillations corresponding to the drops of Hg falling from the capillary. If one connected the maximum current of each drop, a sigmoidal shape would result. The limiting current (the plateau on the sigmoid), called the diffusion current because diffusion is the principal contribution to the flux of electroactive material at this point of the Hg drop life, is related to analyte concentration by the Ilkovic equation:

$$i_d = 708nD^{1/2}m^{2/3}t^{1/6}c$$

Where D is the diffusion coefficient of the analyte in the medium (cm^2/s), n is the number of electrons transferred per mole of analyte, m is the mass flow rate of Hg through the capillary (mg/sec), and t is the drop lifetime is seconds, and c is analyte concentration in mol/cm^3.

LIMITATIONS

There are a number of limitations to the polarography experiment for quantitative analytical measurements. Because the current is continuously measured during the growth of the Hg drop, there is a substantial contribution from capacitive current. As the Hg flows from the capillary end, there is initially a large increase in the surface area. As a consequence, the initial current is dominated by capacitive effects as charging of the rapidly increasing interface occurs. Toward the end of the drop life, there is little change in the surface area which diminishes the contribution of capacitance changes to the total current. At the same time, any redox process which occurs will result in faradaic current that decays approximately as the square root of time (due to the increasing dimensions of the Nernst diffusion layer). The exponential decay of the capacitive current is much more rapid than the decay of the faradaic current; hence, the faradaic current is proportionally larger at the end of the drop life. Unfortunately, this process is complicated by the continuously changing potential that is applied to the working electrode (the Hg drop) throughout the experiment. Because the potential is changing during the drop lifetime (assuming typical experimental parameters of a 2mV/sec scan rate and a 4 sec drop time, the potential can change by 8 mV from the beginning to the end of the drop), the charging of the interface (capacitive current) has a continuous contribution to the total current, even at the end of the drop when the surface area is not rapidly changing. As such, the typical signal to noise of a polarographic experiment allows detection limits of only approximately 10-5 or 10-6 M. Better discrimination against the capacitive current can be obtained using the pulse polarographic techniques.

QUALITATIVE INFORMATION

Qualitative information can also be determined from the half-wave potential of the polarogram (the current vs. potential plot in a polarographic experiment). The value of the half-wave potential is related to the standard potential for the redox reaction being studied.

SPECTROSCOPY

Spectroscopy was originally the study of the interaction between radiation and matter as a function of wavelength (λ). In fact, historically, spectroscopy referred to the use of visible light dispersed according to its wavelength, e.g. by a prism. Later the concept was expanded greatly to comprise any measurement of a quantity as function of either wavelength or frequency. Thus it also can refer to a response to an alternating field or varying frequency (υ). A further extension of the scope of the definition added energy (E) as a variable, once the very close relationship $E=h\upsilon$ for photons was realized. A plot of the response as a function of wavelength—or more commonly frequency—is referred to as a spectrum; see also spectral linewidth.

Spectrometry is the spectroscopic technique used to assess the concentration or amount of a given species. In those cases, the instrument that performs such measurements is a spectrometer or spectrograph.

Spectroscopy/spectrometry is often used in physical and analytical chemistry for the identification of substances through the spectrum emitted from or absorbed by them.

Spectroscopy/spectrometry is also heavily used in astronomy and remote sensing. Most large telescopes have spectrometers, which are used either to measure the chemical composition and physical properties of astronomical objects or to measure their velocities from the Doppler shift of their spectral lines.

CLASSIFICATION OF METHODS

Extremely high resolution spectrum of the Sun showing thousands of elemental absorption lines (fraunhofer lines).

Nature of Excitation Measured

The type of spectroscopy depends on the physical quantity measured. Normally, the quantity that is measured is an intensity, either of energy absorbed or produced.

- Electromagnetic spectroscopy involves interactions of matter with electromagnetic radiation, such as light.
- Electron spectroscopy involves interactions with electron beams. Auger spectroscopy involves inducing the Auger effect with an electron beam. In this case the measurement typically involves the kinetic energy of the electron as variable.
- Mass spectrometry involves the interaction of charged species with magnetic and/or electric fields, giving rise to a mass spectrum. The term "mass spectroscopy" is deprecated, for the technique is primarily a form of measurement, though it does produce a spectrum for observation. This spectrum has the mass m as variable, but the measurement is essentially one of the kinetic energy of the particle.

Acoustic spectroscopy involves the frequency of sound. Dielectric spectroscopy involves the frequency of an external electrical field.

Mechanical spectroscopy involves the frequency of an external mechanical stress, e.g. a torsion applied to a piece of material.

Measurement Process

Most spectroscopic methods are differentiated as either atomic or molecular based on whether or not they apply to atoms or molecules. Along with that distinction, they can be classified on the nature of their interaction:

- Absorption spectroscopy uses the range of the electromagnetic spectra in which a substance absorbs. This includes atomic absorption spectroscopy and various molecular techniques, such as infrared spectroscopy in that region and nuclear magnetic resonance (NMR) spectros-copy in the radio region.
- Emission spectroscopy uses the range of electromagnetic spectra in which a substance radiates (emits). The substance first must absorb energy. This energy can be from a variety of sources, which determines the name of the subsequent emission, like luminescence. Molecular luminescence techniques include spectrofluorimetry.
- Scattering spectroscopy measures the amount of light that a substance scatters at certain wavelengths, incident angles, and polarization angles. The scattering process is much faster than the absorption/emission process. One of the most useful applications of light scattering spectroscopy is Raman spectroscopy.

COMMON TYPES

Absorption

Absorption spectroscopy is a technique in which the power of a beam of light measured before and after interaction with a sample is compared. When performed with tunable diode laser, it is often referred to as Tunable diode laser absorption spectroscopy (TDLAS). It is also often combined with a modulation technique, most often wavelength modulation spectrometry (WMS) and occasionally frequency modulation spectrometry (FMS) in order to reduce the noise in the system.

Fluorescence

Fluorescence spectroscopy uses higher energy photons to excite a sample, which will then emit lower energy photons. This technique has become popular for its biochemical and medical applications, and can be used for confocal microscopy, fluorescence resonance energy transfer, and fluorescence lifetime imaging.

X-ray

When X-rays of sufficient frequency (energy) interact with a substance, inner shell electrons in the atom are excited to outer empty orbitals, or they may be removed completely, ionizing the atom. The inner shell "hole" will then be filled by electrons from outer orbitals. The energy available in this de-excitation process is emitted as radiation (fluorescence) or will remove other less-bound electrons from the atom (Auger effect). The absorption or emission frequencies (energies) are characteristic of the specific atom. In addition, for a specific atom small frequency (energy) variations occur which are characteristic of the chemical bonding. With a suitable apparatus, these characteristic X-ray frequencies or Auger electron energies can be measured. X-ray absorption and emission spectroscopy is used in chemistry and material sciences to determine elemental composition and chemical bonding.

X-ray crystallography is a scattering process; crystalline materials scatter X-rays at well-defined angles. If the wavelength of the incident X-rays is known, this allows calculation of the distances between planes of atoms within the crystal. The intensities of the scattered X-rays give information about the atomic positions and allow the arrangement of the atoms within the crystal structure to be calculated.

Flame

Liquid solution samples are aspirated into a burner or nebulizer/burner combination, desolvated, atomized, and sometimes excited to a higher energy electronic state. The use of a flame during analysis requires fuel and oxidant, typically in the form of gases. Common fuel gases used are acetylene (ethyne) or hydrogen. Common oxidant gases used are oxygen, air, or nitrous oxide. These methods are often capable of analyzing metallic element analytes in the part per million, billion, or possibly lower concentration ranges. Light detectors are needed to detect light with the analysis information coming from the flame.

- *Atomic Emission Spectroscopy:* This method uses flame excitation; atoms are excited from the heat of the flame to emit light. This method commonly uses a total consumption burner with a round burning outlet. A higher temperature flame than atomic absorption spectroscopy (AA) is typically used to produce excitation of analyte atoms. Since analyte atoms are excited by the heat of the flame, no special elemental lamps to shine into the flame are needed. A high resolution polychromator can be used to produce an emission intensity vs. wavelength spectrum over a range of wavelengths showing multiple element excitation lines, meaning multiple elements can be detected in one run. Alternatively, a monochromator can be set at one wavelength to concentrate on analysis of a single element at a certain emission line. Plasma emission spectroscopy is a more modern version of this method. See Flame emission spectroscopy for more details.
- *Atomic Absorption Spectroscopy:* This method commonly uses a pre-burner nebulizer (or nebulizing chamber) to create a sample mist and a slot-shaped burner which gives a longer pathlength flame. The temperature of the flame is low enough that the flame itself does not excite sample atoms from their ground state. The nebulizer and flame are used to desolvate and atomize the sample, but the excitation of the analyte atoms is done by the use of lamps shining through the flame at various wavelengths for each type of analyte. In AA, the amount of light absorbed after going through the flame determines the amount of analyte in the sample. A graphite furnace for heating the sample to desolvate

and atomize is commonly used for greater sensitivity. The graphite furnace method can also analyze some solid or slurry samples. Because of its good sensitivity and selectivity, it is still a commonly used method of analysis for certain trace elements in aqueous (and other liquid) samples.

- *Atomic Fluorescence Spectroscopy:* This method commonly uses a burner with a round burning outlet. The flame is used to solvate and atomize the sample, but a lamp shines light at a specific wavelength into the flame to excite the analyte atoms in the flame. The atoms of certain elements can then fluoresce emitting light in a different direction. The intensity of this fluorescing light is used for quantifying the amount of analyte element in the sample. A graphite furnace can also be used for atomic fluorescence spectroscopy. This method is not as commonly used as atomic absorption or plasma emission spectroscopy.
- *Plasma Emission Spectroscopy:* In some ways similar to flame atomic emission spectroscopy, it has largely replaced it.
  - Direct-current plasma (DCP)

    A direct-current plasma (DCP) is created by an electrical discharge between two electrodes. A plasma support gas is necessary, and Ar is common. Samples can be deposited on one of the electrodes, or if conducting can make up one electrode.
- Glow discharge-optical emission spectrometry (GD-OES)
- Inductively coupled plasma-atomic emission spectrometry (ICP-AES)
- Laser Induced Breakdown Spectroscopy (LIBS) (LIBS), also called Laser-induced plasma spectrometry (LIPS)
- Microwave-induced plasma (MIP)

Spark or arc (emission) spectroscopy

It is used for the analysis of metallic elements in solid samples. For non-conductive materials, a sample is ground with graphite powder to make it conductive. In traditional arc spectroscopy methods, a sample of the solid was commonly ground up and destroyed during analysis. An electric arc or spark is passed through the sample, heating the sample to a high temperature to excite the atoms in it. The excited analyte atoms glow emitting light at various wavelengths which could be detected by common spectroscopic methods. Since the conditions producing the arc emission typically are not controlled quantitatively, the analysis for the elements is qualitative. Nowadays, the spark sources with controlled discharges under an argon atmosphere allow that this method can be considered eminently quantitative, and its use is widely expanded worldwide through production control laboratories of foundries and steel mills.

Visible

Many atoms emit or absorb visible light. In order to obtain a fine line spectrum, the atoms must be in a gas phase. This means that the substance has to be vaporised. The spectrum is studied in absorption or emission. Visible absorption spectroscopy is often combined with UV absorption spectroscopy in UV/Vis spectroscopy. Although this form may be uncommon as the human eye is a similar indicator, it still proves useful when distinguishing colours.

Ultraviolet

All atoms absorb in the Ultraviolet (UV) region because these photons are energetic enough to excite outer electrons. If the frequency is high enough, photoionization takes place. UV spectroscopy is also used in quantifying protein and DNA concentration as well as the ratio of protein to DNA concentration in a solution. Several amino acids usually found in protein, such as tryptophan, absorb light in the 280 nm range and DNA absorbs light in the 260 nm range. For this reason, the ratio of 260/280 nm absorbance is a good general indicator of the relative purity of a solution in terms of these two macromolecules. Reasonable estimates of protein or DNA concentration can also be made this way using Beer's law.

Infrared

Infrared spectroscopy offers the possibility to measure different types of inter atomic bond vibrations at different frequencies. Especially in organic chemistry the analysis of IR absorption spectra shows what type of bonds are present in the sample. It is also an important method for analysing polymers and constituents like fillers. pigments and plasticizers.

Raman

Raman spectroscopy uses the inelastic scattering of light to analyse vibrational and rotational modes of molecules. The resulting 'fingerprints' are an aid to analysis.

Nuclear Magnetic Resonance

Nuclear magnetic resonance spectroscopy analyzes the magnetic properties of certain atomic nuclei to determine different electronic local environments of hydrogen, carbon, or other atoms in an organic compound or other compound. This is used to help determine the structure of the compound.

Mössbauer

Transmission or conversion-electron (CEMS) modes of Mössbauer spectroscopy probe the properties of specific isotope nuclei in different atomic environments by analyzing the resonant absorption of characteristic energy gamma-rays known as the Mössbauer effect.

OTHER TYPES

- Acoustic spectroscopy
- Auger SpectroscopyA method used to study surfaces of materials on a micro-scale. It is often used in connection with electron microscopy.
- Cavity ring down spectroscopy
- Circular Dichroism spectroscopy
- Dielectric spectroscopy
- Force spectroscopy
- *Fourier transform spectroscopy:* An efficient method for processing spectra data obtained using interferometers. Nearly all infrared spectroscopy (Such as FTIR) and Nuclear Magnetic Resonance (NMR) spectroscopy are based on Fourier transforms.
- Fourier transform infrared spectroscopy (FTIR)
- Hadron spectroscopy studies the energy/mass spectrum of hadrons according to spin, parity, and other particle properties. Baryon spectroscopy and meson spectroscopy are both types of hadron spectroscopy.
- Inelastic electron tunnelling spectroscopy (IETS) uses the changes in current due to inelastic electron-vibration interaction at specific energies which can also measure optically forbidden transitions.
- Inelastic neutron scattering like Raman spectroscopy, but with neutrons instead of photons.
- Mechanical spectroscopy involves interactions with macroscopic vibrations, such as phonons. An example is acoustic spectroscopy, involving sound waves.
- Nuclear magnetic resonance (NMR)
- Photoacoustic spectroscopy measures the sound waves produced upon the absorption of radiation.

- Photothermal spectroscopy measures heat evolved upon absorption of radiation.
- Raman optical activity spectroscopy exploits Raman scattering and optical activity effects to reveal detailed information on chiral centers in molecules.
- Terahertz spectroscopy uses wavelengths above infrared spectroscopy and below microwave or millimeter wave measurements.
- Time-resolved spectroscopy is the spectroscopy of matter in situations where the properties are changing with time.
- Thermal infrared spectroscopy measures thermal radiation emitted from materials and surfaces and is used to determine the type of bonds present in a sample as well as their lattice environment. The techniques are widely used by organic chemists, mineralogists, and planetary scientists.

BACKGROUND SUBTRACTION

Background subtraction is a term typically used in spectroscopy when one explains the process of acquiring a background radiation level (or ambient radiation level) and then makes an algorithmic adjustment to the data to obtain qualitative information about any deviations from the background, even when they are an order of magnitude less decipherable than the background itself.

Background subtraction can effect a number of statistical calculations (Continuum, Compton, Bremsstrahlung) leading to improved overall system performance.

CHAPTER - 12

Chromatography

Chromatography (from Greek *chroma,* colour and *graphein* to write) is the collective term for a family of laboratory techniques for the separation of mixtures. It involves passing a mixture dissolved in a "mobile phase" through a *stationary phase,* which separates the analyte to be measured from other molecules in the mixture and allows it to be isolated.

Chromatography may be preparative or analytical. Preparative chromatography seeks to separate the components of a mixture for further use (and is thus a form of purification). Analytical chromatography normally operates with smaller amounts of material and seeks to measure the relative proportions of analytes in a mixture. The two are not mutually exclusive.

HISTORY

The history of chromatography spans from the mid-19th century to the 21st. Chromatography, literally "colour writing", was used—and named—in the first decade of the 20th century, primarily for the separation of plant pigments such as chlorophyll. New forms of chromatography developed in the 1930s and 1940s made the technique useful for a wide range of separation processes.

Some related techniques were developed in the 19th century (and even before), but the first true chromatography is usually attributed to Russian botanist Mikhail Semyonovich Tsvet, who used columns of calcium carbonate for separating plant pigments in the first decade of the 20th century during his research on chlorophyll.

Chromatography began to take its modern form following the work of Archer John Porter Martin and Richard Laurence Millington Synge in the 1940s and 1950s. They laid out the principles and basic techniques of partition chromatography, and their work spurred the rapid development of several lines of chromatography methods: paper chromatography, gas chromatography, and what would become known as high performance liquid chromatography. Since then, the technology has advanced rapidly. Researchers found that the principles underlying Tsvet's chromatography could be applied in many different ways, giving rise to the different varieties of chromatography described below. Simultaneously, advances continually improved the technical performance of chromatography, allowing the separation of increasingly similar molecules.

Chromatography Terms

- The *analyte* is the substance that is to be separated during chromatography.
- *Analytical chromatography* is used to determine the existence and possibly also the concentration of analyte(s) in a sample.
- A *bonded phase* is a stationary phase that is covalently bonded to the support particles or to the inside wall of the column tubing.
- A *chromatogram* is the visual output of the chromatograph. In the case of an optimal separation, different peaks or patterns on the chromatogram correspond to different components of the separated mixture.

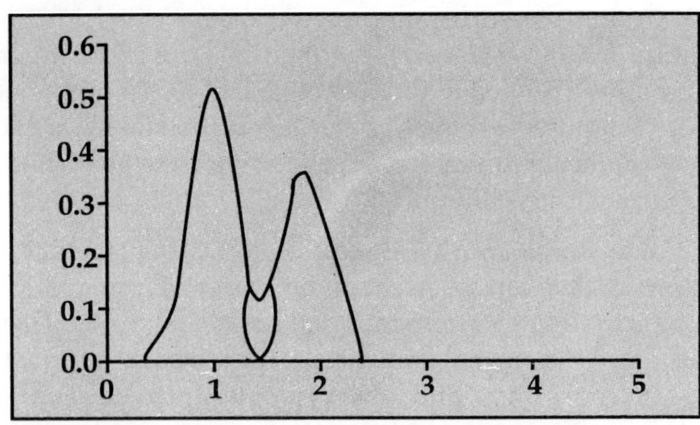

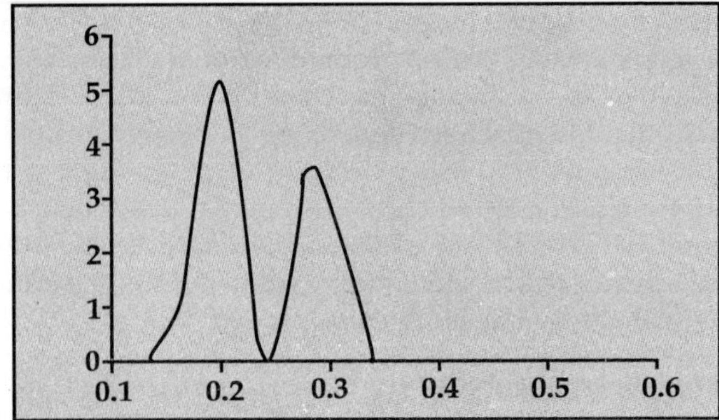

Fig. 12.1: Plotted on the x-axis is the retention time and plotted on the y-axis a signal (for example obtained by a spectrophotometer, mass spectrometer or a variety of other detectors) corresponding to the response created by the analytes exiting the system. In the case of an optimal system the signal is proportional to the concentration of the specific analyte separated.

- A *chromatograph* is equipment that enables a sophisticated separation e.g. gas chromatographic or liquid chromatographic separation.
- *Chromatography* is a physical method of separation in which the components to be separated are distributed

between two phases, one of which is stationary (stationary phase) while the other (the mobile phase) moves in a definite direction.

- The eluent is the mobile phase leaving the column.
- An *immobilized phase* is a stationary phase which is immobilized on the support particles, or on the inner wall of the column tubing.
- The *mobile phase* is the phase which moves in a definite direction. It may be a liquid (LC and CEC), a gas (GC), or a supercritical fluid (supercritical-fluid chromatography, SFC). A better definition: The mobile phase consists of the sample being separated/analyzed and the solvent that moves the sample through the column. In one case of HPLC the solvent consists of a carbonate/bicarbonate solution and the sample is the anions being separated. The mobile phase moves through the chromatography column (the stationary phase) where the sample interacts with the stationary phase and is separated.
- *Preparative chromatography* is used to purify sufficient quantities of a substance for further use, rather than analysis.
- The *retention time* is the characteristic time it takes for a particular analyte to pass through the system (from the column inlet to the detector) under set conditions.
- The *sample* is the matter analysed in chromatography. It may consist of a single component or it may be a mixture of components. When the sample is treated in the course of an analysis, the phase or the phases containing the analytes of interest is/are referred to as the sample whereas everything out of interest separated from the sample before or in the course of the analysis is referred to as waste.

- The *solute* refers to the sample components in partition chromatography.
- The *solvent* refers to any substance capable of solubilizing other substance, and especially the liquid mobile phase in LC.
- The *stationary phase* is the substance which is fixed in place for the chromatography procedure. Examples include the silica layer in Chromatography # Thin layer chromato-graphy.

TECHNIQUES BY CHROMATOGRAPHIC BED SHAPE

Column Chromatography

Column chromatography is a separation technique in which the stationary bed is within a tube. The particles of the solid stationary phase or the support coated with a liquid stationary phase may fill the whole inside volume of the tube (packed column) or be concentrated on or along the inside tube wall leaving an open, unrestricted path for the mobile phase in the middle part of the tube (open tubular column). Differences in rates of movement through the medium are calculated to different retention times of the sample.

In 1978, W.C. Still introduced a modified version of column chromatography called flash column chromatography (flash The technique is very similar to the traditional column chromatography, except for that the solvent is driven through the column by applying positive pressure. This allowed most separations to be performed in less than 20 minutes, with improved separations compared to the old method. Modern flash chromatography systems are sold as pre-packed plastic cartridges, and the solvent is pumped through the cartridge. Systems may also be linked with detectors and fraction collectors providing automation. The introduction of gradient pumps resulted in quicker separations and less solvent usage.

A spreadsheet that assists in the successful development of flash columns has been developed. The spreadsheet estimates the retention volume and band volume of analytes, the fraction numbers expected to contain each analyte, and the resolution between adjacent peaks. This information allows users to select optimal parameters for preparative-scale separations before the flash column itself is attempted.

In expanded bed adsorption, a fluidized bed is used, rather than a solid phase made by a packed bed. This allows omission of initial clearing steps such as centrifugation and filtration, for culture broths or slurries of broken cells.

Planar Chromatography

Planar chromatography is a separation technique in which the stationary phase is present as or on a plane. The plane can be a paper, serving as such or impregnated by a substance as the stationary bed (paper chromatography) or a layer of solid particles spread on a support such as a glass plate (thin layer chromatography). Different compounds in the sample mixture travel different distances according to how strongly they interact with the stationary phase as compared to the mobile phase. The specific Retardation factor (Rf) of each chemical can be used to aid in the identification of an unknown substance.

Paper Chromatography

Paper chromatography is a technique that involves placing a small dot or line of sample solution onto a strip of *chromatography paper*. The paper is placed in a jar containing a shallow layer of solvent and sealed. As the solvent rises through the paper, it meets the sample mixture which starts to travel up the paper with the solvent. This paper is made of cellulose, a polar substance, and the compounds within the mixture travel farther if they are non-polar. More polar substances bond with the cellulose paper more quickly, and therefore do not travel as far.

Thin Layer Chromatography

Thin layer chromatography (TLC) is a widely-employed laboratory technique and is similar to paper chromatography.

However, instead of using a stationary phase of paper, it involves a stationary phase of a thin layer of adsorbent like silica gel, alumina, or cellulose on a flat, inert substrate. Compared to paper, it has the advantage of faster runs, better separations, and the choice between different adsorbents. For even better resolution and to allow for quantitation, high-performance TLC can be used.

DISPLACEMENT CHROMATOGRAPHY

The basic principle of displacement chromatography is: A molecule with a high affinity for the chromatography matrix (the displacer) will compete effectively for binding sites, and thus displace all molecules with lesser affinities. There are distinct differences between displacement and elution chromatography. In elution mode, substances typically emerge from a column in narrow, Gaussian peaks. Wide separation of peaks, preferably to baseline, is desired in order to achieve maximum purification. The speed at which any component of a mixture travels down the column in elution mode depends on many factors. But for two substances to travel at different speeds, and thereby be resolved, there must be substantial differences in some interaction between the biomolecules and the chromatography matrix. Operating parameters are adjusted to maximize the effect of this difference. In many cases, baseline separation of the peaks can be achieved only with gradient elution and low column loadings. Thus, two drawbacks to elution mode chromatography, especially at the preparative scale, are operational complexity, due to gradient solvent pumping, and low throughput, due to low column loadings. Displacement chromatography has advantages over elution chromatography in that components are resolved into consecutive zones of pure substances rather than "peaks". Because the process takes advantage of the nonlinearity of the isotherms, a larger column feed can be separated on a given column with the purified components recovered at significantly higher concentrations.

GAS CHROMATOGRAPHY

Gas chromatography (GC), also sometimes known as Gas-Liquid chromatography, (GLC), is a separation technique in which the mobile phase is a gas. Gas chromatography is always carried out in a column, which is typically "packed" or "capillary".

Gas chromatography (GC) is based on a partition equilibrium of analyte between a solid stationary phase (often a liquid silicone-based material) and a mobile gas (most often Helium). The stationary phase is adhered to the inside of a small-diameter glass tube (a capillary column) or a solid matrix inside a larger metal tube (a packed column). It is widely used in analytical chemistry; though the high temperatures used in GC make it unsuitable for high molecular weight biopolymers or proteins (heat will denature them), frequently encountered in biochemistry, it is well suited for use in the petrochemical, environmental monitoring, and industrial chemical fields. It is also used extensively in chemistry research.

LIQUID CHROMATOGRAPHY

Liquid chromatography (LC) is a separation technique in which the mobile phase is a liquid. Liquid chromatography can be carried out either in a column or a plane. Present day liquid chromatography that generally utilizes very small packing particles and a relatively high pressure is referred to as high performance liquid chromatography (HPLC).

In the HPLC technique, the sample is forced through a column that is packed with irregularly or spherically shaped particles or a porous monolithic layer (stationary phase) by a liquid (mobile phase) at high pressure. HPLC is historically divided into two different sub-classes based on the polarity of the mobile and stationary phases. Technique in which the stationary phase is more polar than the mobile phase (e.g. toluene as the mobile phase, silica as the stationary phase) is called normal phase liquid chromatography (NPLC) and the opposite (e.g. water-methanol mixture as the mobile phase and C18 = octadecylsilyl as the stationary phase) is called reversed phase liquid chromatography (RPLC). Ironically the "normal phase" has fewer applications and RPLC is therefore used considerably more.

Specific techniques which come under this broad heading are listed below. It should also be noted that the following techniques can also be considered fast protein liquid chromatography if no pressure is used to drive the mobile phase through the stationary phase.

AFFINITY CHROMATOGRAPHY

Affinity chromatography is based on selective non-covalent interaction between an analyte and specific molecules. It is very specific, but not very robust. It is often used in biochemistry in the purification of proteins bound to tags. These fusion proteins are labelled with compounds such as His-tags, biotin or antigens, which bind to the stationary phase specifically. After purification, some of these tags are usually removed and the pure protein is obtained.

SUPERCRITICAL FLUID CHROMATOGRAPHY

Supercritical fluid chromatography is a separation technique in which the mobile phase is a fluid above and relatively close to its critical temperature and pressure.

TECHNIQUES BY SEPARATION MECHANISM

Ion Exchange Chromatography

Ion exchange chromatography utilizes ion exchange mechanism to separate analytes. It is usually performed in columns but the mechanism can be benefited also in planar mode. Ion exchange chromatography uses a charged stationary phase to separate charged compounds including amino acids, peptides, and proteins. In conventional methods the stationary phase is an ion exchange resin that carries charged functional groups which interact with oppositely charged groups of the compound to be retained. Ion exchange chromatography is commonly used to purify proteins using FPLC.

Size Exclusion Chromatography

Size exclusion chromatography (SEC) is also known as gel permeation chromatography (GPC) or gel filtration chromato-graphy and separates molecules according to their size (or more accurately according to their hydrodynamic diameter or hydrodynamic volume). Smaller molecules are able to enter the pores of the media and, therefore, take longer to elute, whereas larger molecules are excluded from the pores and elute faster. It is generally a low resolution chromatography technique and thus it is often reserved for the final, "polishing" step of a purification. It is also useful for determining the tertiary structure and quaternary structure of purified proteins, especially since it can be carried out under native solution conditions.

SPECIAL TECHNIQUES

Reversed-phase Chromatography

Reversed-phase chromatography is an elution procedure used in liquid chromatography in which the mobile phase is significantly more polar than the stationary phase.

Two-dimensional Chromatography

In some cases, the chemistry within a given column can be insufficient to separate some analytes. It is possible to direct a series of unresolved peaks onto a second column with different physico-chemical (Chemical classification) properties. Since the mechanism of retention on this new solid support is different from the first dimensional separation, it can be possible to separate compounds that are indistinguishable by one-dimensional chromatography.

Fast Protein Liquid Chromatography

Fast protein liquid chromatography (FPLC) is a term applied to several chromatography techniques which are used to purify proteins. Many of these techniques are identical to those carried out under high performance liquid chromatography, however use of FPLC techniques are typically for preparing large scale batches of a purified product.

Countercurrent Chromatography

Countercurrent chromatography (CCC) is a type of liquid-liquid chromatography, where both the stationary and mobile phases are liquids. It involves mixing a solution of liquids, allowing them to settle into layers and then separating the layers.

Chiral Chromatography

Chiral chromatography involves the separation of stereoisomers. In the case of enantiomers, these have no chemical or physical differences apart from being three dimensional mirror images. Conventional chromatography or other separation processes are incapable of separating them. To enable chiral separations to take place, either the mobile phase or the stationary phase must themselves be made chiral, giving differing affinities between the analytes. Chiral chromatography HPLC columns (with a chiral stationary phase) in both normal and reversed phase are commercially available.

ELECTROPHORESIS

Electrophoresis is the best-known electrokinetic phenomenon. It was discovered by Reuss in 1807 He observed that clay particles dispersed in water migrate under influence of an applied electric field. There are detailed descriptions of Electrophoresis in many books on Colloid and Interface Science There is an IUPAC Technical Report prepared by a group of well known experts on the electrokinetic phenomena. Generally, electrophoresis is the motion of dispersed particles relative to a fluid under the influence of an electric field that is space uniform. Alternatively, similar motion in a space non-uniform electric field is called dielectrophoresis.

Electrophoresis occurs because particles dispersed in a fluid almost always carry an electric surface charge. An electric field exerts electrostatic Coulomb force on the particles through these charges. Recent molecular dynamics simulations, though, suggest that surface charge is not always necessary for electrophoresis and that even neutral particles can show electrophoresis due to the specific molecular structure of water at the interface.

The electrostatic Coulomb force exerted on a surface charge is reduced by an opposing force which is electrostatic as well. According to double layer theory, all surface charges in fluids are screened by a diffuse layer. This diffuse layer has the same absolute charge value, but with opposite sign from the surface charge. The electric field induces force on the diffuse layer, as well as on the surface charge. The total value of this force equals to the first mentioned force, but it is oppositely directed. However, only part of this force is applied to the particle. It is actually applied to the ions in the diffuse layer. These ions are at some distance from the particle surface. They transfer part of this electrostatic force to the particle surface through viscous stress. This part of the force that is applied to the particle body is called electrophoretic retardation force.

There is one more electric force, which is associated with deviation of the double layer from spherical symmetry and surface conductivity due to the excess ions in the diffuse layer. This force is called the electrophoretic relaxation force.

All these forces are balanced with hydrodynamic friction, which affects all bodies moving in viscous fluids with low Reynolds number. The speed of this motion v is proportional to the electric field strength E if the field is not too strong. Using this assumption makes possible the introduction of electrophoretic mobility μe as coefficient of proportionality between particle speed and electric field strength:

$$\mu_e = \frac{\upsilon}{E}$$

Multiple theories were developed during 20th century for calculating this parameter.

THEORY

The most known and widely used theory of electrophoresis was developed by Smoluchowski in 1903.

$$\mu_e = \frac{\varepsilon\varepsilon_0\xi}{\eta}$$

where e is the dielectric constant of the dispersion medium, e_0 is the permittivity of free space ($C^2\ N^{-1}\ m^{-2}$), is dynamic viscosity of the dispersion medium (Pa s), and x is zeta potential (i.e., the electrokinetic potential of the slipping plane in the double layer).

Smoluchowski theory is very powerful because it works for dispersed particles of any shape and any concentration, when it is valid. Unfortunately, it has limitations of its validity. It follows, for instance, from the fact that it does not include Debye length k^{-1}. However, Debye length must be important for electrophoresis, as follows immediately from the Figure on the right. Increasing thickness of the DL leads to removing point of retardation force further from the particle surface. The thicker DL, the smaller retardation force must be.

Detailed theoretical analysis proved that Smoluchowski theory is valid only for sufficiently thin DL, when Debye length is much smaller than particle radius *a*:

$$a\kappa \gg 1$$

This model of "thin Double Layer" offers tremendous simplifications not only for electrophoresis theory but for many other electrokinetic theories. This model is valid for most aqueous systems because the Debye length is only a few nanometers there. It breaks only for nano-colloids in solution with ionic strength close to water.

Smoluchowski theory also neglects contribution of surface conductivity. This is expressed in modern theory as condition of small Dukhin number

$$Du \gg 1$$

Creation of electrophoretic theory with wider range of validity was a purpose of many studies during 20th century.

One of the most known considers an opposite asymptotic case when Debye length is larger than particle radius:

$$ka < 1$$

It is called the "thick Double Layer" model. Corresponding electrophoretic theory was created by Huckel in 1924 It yields the following equation for electrophoretic mobility:

$$\mu_e = \frac{2\varepsilon\varepsilon_0\xi}{3\eta}$$

This model can be useful for some nano-colloids and non-polar fluids, where Debye length is much larger.

There are several analytical theories that incorporate surface conductivity and eliminate restriction of the small Dukhin number. Early pioneering work in that direction dates back to Overbeek and Booth.

Modern, rigorous theories that are valid for any Zeta potential and often any *ka*, stem mostly from the Ukrainian

(Dukhin, Shilov and others) and Australian (O'Brien, White, Hunter and others) Schools.

Historically the first one was Dukhin-Semenikhin theory Similar theory was created 10 years later by O'Brien and Hunter. Assuming thin Double Layer, these theories would yield results that are very close to the numerical solution provided by O'Brien and White.

MICROSCOPY

Microscopy is the technical field of using microscopes to view samples or objects. There are three well-known branches of microscopy, optical, electron and scanning probe microscopy.

Optical and electron microscopy involve the diffraction, reflection, or refraction of electromagnetic radiation interacting with the subject of study, and the subsequent collection of this scattered radiation in order to build up an image. This process may be carried out by wide-field irradiation of the sample (for example standard light microscopy and transmission electron microscopy) or by scanning of a fine beam over the sample (for example confocal laser scanning microscopy and scanning electron microscopy). Scanning probe microscopy involves the interaction of a scanning probe with the surface or object of interest. The development of microscopy revolutionized biology and remains an essential tool in that science, along with many others including materials science and numerous engineering disciplines.

OPTICAL MICROSCOPY

Optical or light microscopy involves passing visible light transmitted through or reflected from the sample through a single or multiple lenses to allow a magnified view of the sample. The resulting image can be detected directly by the eye, imaged on a photographic plate or captured digitally. The single lens with its attachments, or the system of lenses and imaging equipment, along with the appropriate lighting equipment, sample stage and support, makes up the basic light microscope. The most recent development is the digital microscope which uses a CCD camera to focus on the exhibit of interest. The image is shown on a computer screen since the camera is attached to it via a USB port, so eye-pieces are unnecessary.

Limitations

Limitations of standard optical microscopy (bright field microscopy) lie in three areas:

- The technique can only image dark or strongly refracting objects effectively.
- Diffraction limits resolution to approximately 0.2 micrometre.
- Out of focus light from points outside the focal plane reduces image clarity.

Live cells in particular generally lack sufficient contrast to be studied successfully, internal structures of the cell are colourless and transparent. The most common way to increase contrast is to stain the different structures with selective dyes, but this involves killing and fixing the sample. Staining may also introduce artifacts, apparent structural details that are caused by the processing of the specimen and are thus not a legitimate feature of the specimen.

These limitations have all been overcome to some extent by specific microscopy techniques which can non-invasively increase the contrast of the image. In general, these techniques make use of differences in the refractive index of cell structures. It is comparable to looking through a glass window: you (bright field microscopy) don't see the glass but merely the dirt on the glass. There is however a difference as glass is a denser material, and this creates a difference in phase of the light passing through. The human eye is not sensitive to this difference in phase but clever optical solutions have been thought out to change this difference in phase into a difference in amplitude (light intensity).

TECHNIQUES

Bright Field

Bright field microscopy is the simplest of all the light microscopy techniques. Sample illumination is via transmitted white light, i.e. illuminated from below and observed from above. Limitations include low contrast of most biological samples and low apparent resolution due to the blur of out of focus material. The simplicity of the technique and the minimal sample preparation required are significant advantages.

Oblique Illumination

The use of oblique (from the side) illumination gives the image a 3-dimensional appearance and can highlight otherwise invisible features. A more recent technique based on this method is *Hoffmann's modulation contrast,* a system found on inverted microscopes for use in cell culture. Oblique illumination suffers from the same limitations as bright field microscopy (low contrast of many biological samples; low apparent resolution due to out of focus objects), but may highlight otherwise invisible structures.

Dark field

Dark field microscopy is a technique for improving the contrast of unstained, transparent specimens Dark field illumination uses a carefully aligned light source to minimize the quantity of directly-transmitted (unscattered) light entering the image plane, collecting only the light scattered by the sample. Darkfield can dramatically improve image contrast—especially of transparent objects – while requiring little equipment setup or sample preparation. However, the technique does suffer from low light intensity in final image of many biological samples, and continues to be affected by low apparent resolution.

Rheinberg illumination is a special variant of dark field illumination in which transparent, colored filters are inserted just before the condenser so that light rays at high aperture are differently colored than those at low aperture (i.e. the background to the specimen may be blue while the object appears self-luminous yellow). Other color combinations are possible but their effectiveness is quite variable

Dispersion Staining

Dispersion staining is an optical technique that results in a colored image of a colorless object. This is an optical staining technique and requires no stains or dyes to produce a color effect. There are five different microscope configurations used in the broader technique of dispersion staining. They include brightfield Becke' line, oblique, darkfield, phase contrast, and objective stop dispersion staining.

Phase Contrast

More sophisticated techniques will show proportional differences in optical density. Phase contrast is a widely used technique that shows differences in refractive index as difference in contrast. It was developed by the Dutch physicist Frits Zernike in the 1930s (for which he was awarded the Nobel Prize in 1953). The nucleus in a cell for example will show up darkly against the surrounding cytoplasm. Contrast is excellent; however it is not for use with thick objects. Frequently, a halo is formed even around small objects, which obscures detail. The system consists of a circular annulus in the condenser which produces a cone of light. This cone is superimposed on a similar sized ring within the phase-objective. Every objective has a different size ring, so for every objective another condenser setting has to be chosen. The ring in the objective has special optical properties: it first of all reduces the direct light in intensity, but more importantly, it creates an artificial phase difference of about a quarter wavelength. As the physical properties of this direct light have changed, interference with the diffracted light occurs, resulting in the phase contrast image.

Differential Interference Contrast

Superior and much more expensive is the use of interference contrast. Differences in optical density will show up as differences in relief. A nucleus within a cell will actually show up as a globule in the most often used differential interference contrast system according to Georges Nomarski. However, it has to be kept in mind that this is an *optical effect*, and the relief does not necessarily resemble the true shape! Contrast is very good and the condenser aperture can be used fully open, thereby reducing the depth of field and maximizing resolution.

The system consists of a special prism (Nomarski prism, Wollaston prism) in the condenser that splits light in an ordinary and an extraordinary beam. The spatial difference between the two beams is minimal (less than the maximum resolution of the objective). After passage through the specimen, the beams are reunited by a similar prism in the objective.

In a homogeneous specimen, there is no difference between the two beams, and no contrast is being generated. However, near a refractive boundary (say a nucleus within the cytoplasm), the difference between the ordinary and the extraordinary beam will generate a relief in the image. Differential interference contrast requires a polarized light source to function; two polarizing filters have to be fitted in the light path, one below the condenser (the polarizer), and the other above the objective (the analyzer).

Note: In cases where the optical design of a microscope produces an appreciable lateral separation of the two beams we have the case of classical interference microscopy, which does not result in relief images, but can nevertheless be used for the quantitative determination of mass-thicknesses of microscopic objects.

Fluorescence

When certain compounds are illuminated with high energy light, they then emit light of a different, lower frequency. This effect is known as fluorescence. Often specimens show their own characteristic autofluorescence image, based on their chemical makeup.

This method is of critical importance in the modern life sciences, as it can be extremely sensitive, allowing the detection of single molecules. Many different fluorescent dyes can be used to stain different structures or chemical compounds. One particularly powerful method is the combination of antibodies coupled to a fluorochrome as in immunostaining. Examples of commonly used fluorochromes are fluorescein or rhodamine. The antibodies can be made tailored specifically for a chemical compound. For example, one strategy often in use is the artificial production of proteins, based on the genetic code (DNA). These proteins can then be used to immunize rabbits, which then form antibodies which bind to the protein. The antibodies are then coupled chemically to a fluorochrome and then used to trace the proteins in the cells under study.

Highly-efficient fluorescent proteins such as the green fluorescent protein (GFP) have been developed using the molecular biology technique of gene fusion, a process which links the expression of the fluorescent compound to that of the target protein. "Nikon MicroscopyU: Introduction to Fluorescent Proteins". This combined fluorescent protein is generally non-toxic to the organism and rarely interferes with the function of the protein under study. Genetically modified cells or organisms directly express the fluorescently-tagged proteins, which enables the study of the function of the original protein in vivo.

Since fluorescence emission differs in wavelength (color) from the excitation light, a fluorescent image ideally only shows the structure of interest that was labeled with the fluorescent dye. This high specificity led to the widespread use of fluorescence light microscopy in biomedical research. Different fluorescent dyes can be used to stain different biological structures, which can then be detected simultaneously, while still being specific due to the individual color of the dye.

To block the excitation light from reaching the observer or the detector, filter sets of high quality are needed. These typically consist of an excitation filter selecting the range of excitation wavelengths, a dichroic mirror, and an emission filter blocking the excitation light. Most fluorescence microscopes are operated in the Epi-illumination mode (illumination and detection from one side of the sample) to further decrease the amount of excitation light entering the detector.

Confocal Laser Scanning Microscopy

Confocal laser scanning (CLSM) generates the image by a completely different way than the normal visual bright field microscope. It gives slightly higher resolution, but most importantly it provides optical sectioning without disturbing out-of-focus light degrading the image. Therefore it provides sharper images of 3D objects. This is often used in conjunction with fluorescence microscopy.

Deconvolution

Fluorescence microscopy is extremely powerful due to its ability to show specifically labeled structures within a complex environment but also because of its inherent ability to provide three dimensional information of biological structures. Unfortunately this information is blurred by the fact, that upon illumination all fluorescently labeled structures emit light no matter if they are in focus or not.

This means, that an image of a certain structure is always blurred by the contribution of light from structures which are out of focus. This phenomenon becomes apparent as a loss of contrast especially when using objectives with a high resolving power, typically oil immersion objectives with a high numerical aperture.

Fortunately though, this phenomenon is not caused by random processes such as light scattering but can be relatively well defined by the optical properties of the image formation in the microscope imaging system. If one considers a small fluorescent light source (essentially a bright spot), light coming from this spot spreads out the further out of focus one is. Under ideal conditions this produces a sort of "hourglass" shape of this point source in the third (axial) dimension. This shape is called the point spread function (PSF) of the microscope imaging system. Since any fluorescence image is made up of a large number of such small fluorescent light sources the image is said to be "convolved by the point spread function".

Knowing this point spread function means, that it is possible to reverse this process to a certain extent by computer based methods commonly known as deconvolution microscopy. There are various algorithms available for 2D or 3D Deconvolution. They can be roughly classified in *non restorative* and *restorative* methods. While the non restorative methods can improve contrast by removing out of focus light from focal planes, only the restorative methods can actually reassign light to it proper place of origin. This can be an advantage over other types of 3D microscopy such as confocal microscopy, because light is not thrown away but reused. For 3D deconvolution one typically provides a series of images derived from different focal planes (called a Z-stack) plus the knowledge of the PSF which can be either derived experimentally or theoretically from knowing all contributing parameters of the microscope.

Sub-diffraction Techniques

It is well known that there is a spatial limit to which light can focus: approximately half of the wavelength of the light you are using. But this is not a true barrier, because this diffraction limit is only true in the far-field and localization precision can be increased with many photons and careful analysis (although two objects still cannot be resolved); and like the sound barrier, the diffraction barrier is breakable. This section explores some approaches to imaging objects smaller than ~250 nm. Most of the following information was gathered (with permission) from a chemistry blog's review of sub-diffraction microscopy techniques Part I and Part II. For a review, see also reference.

Near-field Scanning

Near-field scanning is also called NSOM. Probably the most conceptual way to break the diffraction barrier is to use a light source and/or a detector that is itself nanometer in scale. Diffraction as we know it is truly a far-field effect: the light from an aperture is the Fourier transform of the aperture in the far-field. But in the near-field, all of this is not necessarily the case. Near-field scanning optical microscopy (NSOM) forces light through the tiny tip of a pulled fiber—and the aperture can be on the order of tens of nanometers. When the tip is brought to nanometers away from a molecule, the resolution is not limited by diffraction but by the size of the tip aperture (because only that one molecule will see the light coming out of the tip). An image can be built by a raster scan of the tip over the surface to create an image.

The main down-side to NSOM is the limited number of photons you can force out a tiny tip, and the minuscule collection efficiency (if you are trying to collect fluorescence in the near-field). Other techniques such as ANSOM (see below) try to avoid this drawback.

Local Enhancement/ANSOM/bowties

Instead of forcing photons down a tiny tip, some techniques create a local bright spot in an otherwise diffraction-limited spot. ANSOM is apertureless NSOM: it uses a tip very close to a fluorophore to enhance the local electric field the fluorophore sees Basically, the ANSOM tip is like a lightning rod which creates a hot spot of light.

Bowtie nanoantennas have been used to greatly and reproducibly enhance the electric field in the nanometer gap between the tips two gold triangles. Again, the point is to enhance a very small region of a diffraction-limited spot, thus improving the mismatch between light and nanoscale objects—and breaking the diffraction barrier

Stimulated Emission Depletion

Stefan Hell at the Max Planck Institute for Biophysical Chemistry - Goettingen (Germany) developed STED microscopy (stimulated emission depletion), which uses two laser pulses. The first pulse is a diffraction-limited spot that is tuned to the absorption wavelength, so excites any fluorophores in that region; an immediate second pulse is red-shifted to the emission wavelength and stimulates emission back to the ground state before, thus depleting the excited state of any fluorophores in this depletion pulse. The trick is that the depletion pulse goes through a phase modulator that makes the pulse illuminate the sample in the shape of a donut, *so the outer part of the diffraction limited spot is depleted and the small center can still fluoresce.* By saturating the depletion pulse, the center of the donut gets smaller and smaller until they can get resolution of tens of nanometers.

This technique also requires a raster scan like NSOM and standard confocal laser scanning microscopy.

Fitting the Point-spread Function

Fitting the point-spread function is also called PSF. The methods above (and below) use experimental techniques to circumvent the diffraction barrier, but one can also use crafty analysis to increase the ability to know where a nanoscale object is located. The image of a point source on a charge-coupled device camera is called a point-spread function (PSF), which is limited by diffraction to be no less than approximately half the wavelength of the light. But it is possible to simply fit that PSF with a Gaussian to locate the center of the PSF—and thus the location of the fluorophore. The precision by which this technique can locate the center depends on the number of photons collected (as well as the CCD pixel size and other factors) Regardless, groups like the Selvin lab and many others have employed this analysis to localize single fluorophores to a few nanometers. This, of course, requires careful measurements and collecting *many* photons.

Palm, Storm

What fitting a PSF is to localization, photo-activated localization microscopy (PALM) is to "resolution"—this term is here used loosely to mean measuring the distance between objects, not true optical resolution. Eric Betzig and colleagues developed PALM Xiaowei Zhuang at Harvard used a similar techniques and calls it STORM: stochastic optical reconstruction microscopy. Sam Hess at University of Maine developed the technique simultaneously. The basic premise of both techniques is to fill the imaging area with many dark fluorophores that can be photoactivated into a fluorescing state by a flash of light. Because photoactivation is stochastic, only a few, well separated molecules "turn on." Then Gaussians are fit to their PSFs to high precision. After the few bright dots photobleach, another flash of the photoactivating light activates random fluorophores again and the PSFs are fit of these different well spaced objects. This process is repeated many times, building up an image molecule-by-molecule; and because the molecules were localized at different times, the "resolution" of the final image can be much higher than that limited by diffraction.

The major problem with these techniques is that to get these beautiful pictures, it takes on the order of hours to collect the data. This is certainly not the technique to study dynamics.

Structured Illumination

There is also the wide-field structured-illumination (SI) approach to breaking the diffraction limit of light SI—or patterned illumination—relies on both specific microscopy protocols and extensive software analysis post-exposure. But, because SI is a wide-field technique, it is usually able to capture images at a higher rate than confocal-based schemes like STED. (This is only a generalization, because SI isn't actually super fast. I'm sure someone could make STED fast and SI slow!) The main concept of SI is to illuminate a sample with patterned light and increase the resolution by measuring the fringes in the Moiré pattern (from the interference of the illumination pattern and the sample). "Otherwise-unobservable sample information can be deduced from the fringes and computationally restored."

SI enhances spatial resolution by collecting information from frequency space outside the observable region. This process is done in reciprocal space: the Fourier transform (FT) of an SI image contains superimposed additional information from different areas of reciprocal space; with several frames with the illumination shifted by some phase, it is possible to computationally separate and reconstruct the FT image, which has much more resolution information. The reverse FT returns the reconstructed image to a super-resolution image.

But this only enhances the resolution by a factor of 2 (because the SI pattern cannot be focused to anything smaller than half the wavelength of the excitation light). To further increase the resolution, you can introduce *nonlinearities,* which show up as higher-order harmonics in the FT. In reference Gustafsson uses saturation of the fluorescent sample as the nonlinear effect. A sinusoidal saturating excitation beam produces the distorted fluorescence intensity pattern in the emission. This nonpolynomial nonlinearity yields a series of higher-order harmonics in the FT.

Each higher-order harmonic in the FT allows another set of images that can be used to reconstruct a larger area in reciprocal space, and thus a higher resolution. In this case, Gustafsson achieves less than 50-nm resolving power, more than five times that of the microscope in its normal configuration.

The main problems with SI are that, in this incarnation, saturating excitation powers cause more photodamage and lower fluorophore photostability, and sample drift must be kept to below the resolving distance. The former limitation might be solved by using a different nonlinearity (such as stimulated emission depletion or reversible photoactivation, both of which are used in other sub-diffraction imaging schemes); the latter limits live-cell imaging and may require faster frame rates or the use of some fiduciary markers for drift subtraction. Nevertheless, SI is certainly a strong contender for further application in the field of super-resolution microscopy.

Localization Microscopy/Spatially Structured Illumination

Around 1995, Christoph Cremer commenced with the development of a light microscopic process, which achieved a substantially improved size resolution of cellular nanostructures stained with a fluorescent marker. This time he employed the principle of wide field microscopy combined with structured laser illumination (spatially modulated illumination, SM Currently, a size resolution of 30-40 nm (approximately 1/16 – 1/13 of the wave length used) is being achieved. In addition, this technology is no longer subjected to the speed limitations of the focusing microscopy so that it becomes possible to undertake 3D analyses of whole cells within short observation times (at the moment around a few seconds). Also since around 1995, Christoph Cremer developed and realized new fluorescence based wide field microscopy approaches which had as their goal the improvement of the effective optical resolution (in terms of the smallest detectable distance between two localized objects) down to a fraction of the conventional resolution (spectral precision distance/position determination microscopy, SPDM). Combining SPDM and SMI, known as Vertico-SMI microscopy Christoph Cremer can currently achieve a resolution of approx. 10 nm in 2D and 40 nm in 3D in wide field images of whole living cells. Widefield 3D "nanoimages" of whole living cells currently still take about two minutes, but work to reduce this further is currently under way. Vertico-SMI is currently the fastest optical 3D nanoscope for the three dimensional structural analysis of whole cells world-wide.

Extensions

Most modern instruments provide simple solutions for micro-photography and image recording electronically. However such capabilities are not always present and the more experienced microscopist will, in many cases, still prefer a hand drawn image rather than a photograph. This is because a microscopist with knowledge of the subject can accurately convert a three dimensional image into a precise two dimensional drawing. In a photograph or other image capture system however, only one thin plane is ever in good focus.

The creation of careful and accurate micrographs requires a microscopical technique using a monocular eyepiece. It is essential that both eyes are open and that the eye that is not observing down the microscope is instead concentrated on a sheet of paper on the bench besides the microscope. With practice, and without moving the head or eyes, it is possible to accurately record the observed details by tracing round the observed shapes by simultaneously "seeing" the pencil point in the microscopical image.

Practicing this technique also establishes good general microscopical technique. It is always less tiring to observe with the microscope focused so that the image is seen at infinity and with both eyes open at all times.

Other Enhancements

X-ray

As resolution depends on the wavelength of the light. Electron microscopy has been developed since the 1930s that use electron beams instead of light. Because of the much lower wavelength of the electron beam, resolution is far higher.

Though less common, X-ray microscopy has also been developed since the late 1940s. The resolution of X-ray microscopy lies between that of light microscopy and the electron microscopy.

ELECTRON MICROSCOPY

For light microscopy the wavelength of the light limits the resolution to around 0.2 micrometers. In order to gain higher resolution, the use of an electron beam with a far smaller wavelength is used in electron microscopes.

Transmission electron microscopy (TEM) is principally quite similar to the compound light microscope, by sending an electron beam through a very thin slice of the specimen. The resolution limit nowadays (2005) is around 0.05 nanometer.

Scanning electron microscopy (SEM) visualizes details on the surfaces of cells and particles and gives a very nice 3D view. It gives results much like the stereo light microscope and akin to that its most useful magnification is in the lower range than that of the transmission electron microscope.

Atomic de Broglie

The *atomic de Broglie microscope* is an imaging system which is expected to provide resolution at the nanometer scale using neutral He atoms as probe particles. Such a device could provide the resolution at nanometer scale and be absolutely non-destructive, but it is not developed so well as optical microscope or an electron microscope.

SCANNING PROBE MICROSCOPY

This is a sub-diffraction technique. Examples of scanning probe microscopes are the atomic force microscope (AFM), the Scanning tunneling microscope and the photonic force microscope. All such methods imply a solid probe tip in the vicinity (near field) of an object, which is supposed to be almost flat.

Ultrasonic Force

Ultrasonic Force Microscopy (UFM) has been developed in order to improve the details and image contrast on "flat" areas of interest where the AFM images are limited in contrast. The combination of AFM-UFM allows a near field acoustic microscopic image to be generated. The AFM tip is used to detect the ultrasonic waves and overcomes the limitation of wavelength that occurs in acoustic microscopy. By using the elastic changes under the AFM tip, an image of much greater detail than the AFM topography can be generated.

Ultrasonic force microscopy allows the local mapping of elasticity in atomic force microscopy by the application of ultrasonic vibration to the cantilever or sample. In an attempt to analyse the results of ultrasonic force microscopy in a quantitative fashion, a force-distance curve measurement is done with ultrasonic vibration applied to the cantilever base, and the results are compared with a model of the cantilever dynamics and tip-sample interaction based on the finite-difference technique.

INFRARED MICROSCOPY

The term *infrared microscope* covers two main types of diffraction-limited microscopy. The first provides optical visualization plus IR spectroscopic data collection. The second (more recent and more advanced) technique employs *focal plane array detection* for infrared chemical imaging, where the image contrast is determined by the response of individual sample regions to particular IR wavelengths selected by the user.

IR versions of sub-diffraction microscopy exist also. These include IR NSOM and photothermal microspectroscopy.

DIGITAL HOLOGRAPHIC MICROSCOPY

In digital holographic microscopy (DHM), interfering wave-fronts from a coherent light-source are recorded on a sensor and the image digitally reconstructed by a computer. The image yielded provides a quantitative measurement of the optical thickness of the specimen. DHM can be used with many different optical set-ups. In reflecting DHM, the sensor is positioned on the same side of the specimen as the light source. In transmitting DHM, the sensor and the light source are positioned on opposite sides of the specimen.

One unique feature of DHM is the ability to adjust focus after the image is recorded, since all focus planes are recorded simultaneously by the hologram.

DIGITAL PATHOLOGY (VIRTUAL MICROSCOPY)

Digital Pathology is an image-based information environment enabled by computer technology that allows for the management of information generated from a digital slide. Digital pathology is enabled in part by virtual microscopy, which is the practice of converting glass slides into digital slides that can be viewed, managed, and analyzed.

AMATEUR MICROSCOPY

Amateur Microscopy is the investigation and observation of biological and non-biological specimens for recreational purposes. Collectors of minerals, insects, seashells and plants may use microscopes as tools to uncover features that help them classify their collected items. Other amateurs may be interested in observing the life found in pond water and of other samples. Microscopes may also prove useful for the water quality assessment for people that keep a home aquarium. Photographic documentation and drawing of the microscopic images are additional tasks that augment the spectrum of tasks of the amateur. There are even competitions for photomicrograph art. Participants of this pastime may either use commercially prepared microscopic slides or may engage in the task of specimen preparation.

While microscopy is a central tool in the documentation of biological specimens, it is generally insufficient to justify the description of a new species based on microscopic investigations alone. Often genetic and biochemical tests are necessary to confirm the discovery of a new species. A laboratory and access to academic literature is a necessity, which is specialized and generally not available to amateurs. There is however one huge advantage that amateurs have above professionals: time to explore their surroundings. Often, advanced amateurs team up with professionals to validate their findings and (possibly) describe new species.

In the late 1800s amateur microscopy became a popular hobby in the United States and Europe. Several 'professional amateurs' were being paid for their sampling trips and microscopic explorations by philanthropists, to keep them amused on the Sunday afternoon (e.g. the diatom specialist A. Grunow, being paid by (among others) a Belgian industrialist). Professor John Phin published "Practical Hints on the Selection and Use of the Microscope (Second Edition, 1878)," and was also the editor of the "American Journal of Microscopy."

In 1995, a loose group of amateur microscopists, drawn from several organizations in the UK and USA, founded a site for microscopy based on the knowledge and input of amateur (perhaps better referred to as 'enthusiast') microscopists. This was historically the first attempt to establish 'amateur' microscopy as a serious subject in the then emerging new media of the Internet. Today, it remains as a powerful established international resource for all ages, to input their findings and share information. It is a non-profit making web presence dedicated to the pursuit of science and understanding of the small-scale world

THERMOGRAVIMETRIC ANALYSIS

Thermogravimetric Analysis or TGA is a type of testing that is performed on samples to determine changes in weight in relation to change in temperature. Such analysis relies on a high degree of precision in three measurements: weight, temperature, and temperature change. As many weight loss curves look similar, the weight loss curve may require transformation before results may be interpreted. A derivative weight loss curve can be used to tell the point at which weight loss is most apparent. Again, interpretation is limited without further modifications and deconvolution of the overlapping peaks may be required.

TGA is commonly employed in research and testing to determine characteristics of materials such as polymers, to determine degradation temperatures, absorbed moisture content of materials, the level of inorganic and organic components in materials, decomposition points of explosives, and solvent residues. It is also often used to estimate the corrosion kinetics in high temperature oxidation.

ANALYZER

The analyzer usually consists of a high-precision balance with a pan (generally platinum) loaded with the sample. The pan is placed in a small electrically heated oven with a thermocouple to accurately measure the temperature. The atmosphere may be purged with an inert gas to prevent oxidation or other undesired reactions. A computer is used to control the instrument.

Analysis is carried out by raising the temperature gradually and plotting weight against temperature. The temperature in many testing methods routinely reaches 1000° C or greater, but the oven is so greatly insulated that an operator would not be aware of any change in temperature even if standing directly in front of the device. After the data is obtained, curve smoothing and other operations may be done such as to find the exact points of inflection.

A method known as hi-res TGA is often employed to obtain greater accuracy in areas where the derivative curve peaks. In this method, temperature increase slows as weight loss increases. This is done so that the exact temperature at which a peak occurs can be more accurately identified. Several modern TGA devices can vent burnoff to a fourier-transform infrared spectro-photometer to analyze composition.

CHAPTER - 13

Crystallography

Crystallography is the experimental science of determining the arrangement of atoms in solids. In older usage, it is the scientific study of crystals. The word "crystallography" is derived from the Greek words crystallon = cold drop/frozen drop, with its meaning extending to all solids with some degree of transparency, and graphein = write.

Before the development of X-ray diffraction crystallography the study of crystals was based on the geometry of the crystals. This involves measuring the angles of crystal faces relative to theoretical reference axes (crystallographic axes), and establishing the symmetry of the crystal in question. The former is carried out using a goniometer. The position in 3D space of each crystal face is plotted on a stereographic net, e.g. Wulff net or Lambert net. In fact, the pole to each face is plotted on the net. Each point is labelled with its Miller index. The final plot allows the symmetry of the crystal to be established.

Crystallographic methods now depend on the analysis of the diffraction patterns that emerge from a sample that is targeted by a beam of some type. The beam is not always electromagnetic radiation, even though X-rays are the most common choice. For some purposes electrons or neutrons are used, which is possible due to the wave properties of the particles. Crystallographers often explicitly state the type of illumination used when referring to a method, as with the terms X-ray diffraction, neutron diffraction and electron diffraction.

These three types of radiation interact with the specimen in different ways. X-rays interact with the spatial distribution of the valence electrons, while electrons are charged particles and therefore feel the total charge distribution of both the atomic nuclei and the surrounding electrons. Neutrons are scattered by the atomic nuclei through the strong nuclear forces, but in addition, the magnetic moment of neutrons is non-zero. They are therefore also scattered by magnetic fields. When neutrons are scattered from hydrogen-containing materials, they produce diffraction patterns with high noise levels. However, the material can sometimes be treated to substitute hydrogen for deuterium. Because of these different forms of interaction, the three types of radiation are suitable for different crystallographic studies.

THEORY

An image of a small object is usually generated by using a lens to focus the illuminating radiation, as is done with the rays of the visible spectrum in light microscopy. However, the wavelength of visible light (about 4000 to 7000 Angstroms) is three orders of magnitude longer then the length of typical atomic bonds and atoms themselves (about 1 to 2 Angstroms). Therefore, obtaining information about the spatial arrangement of atoms requires the use of radiation with shorter wavelengths, such as X-rays. Employing shorter wavelengths implied abandoning microscopy and true imaging, however, because there exists no material from which a lens capable of focusing this type of radiation can be created. (That said, scientists have had some success focusing X-rays with microscopic Fresnel zone plates made from gold, and by critical-angle reflection inside long tapered capillaries Diffracted x-ray beams cannot be focussed to produce images, so the sample structure must be reconstructed from the diffraction pattern. Sharp features in the diffraction pattern arise from periodic, repeating structure in the sample, which are often very strong due to coherent reflection of many photons from many regularly spaced instances of similar structure, while non-periodic components of the structure result in diffuse (and usually weak) diffraction features.

Because of their highly ordered and repetitive structure, crystals give diffraction patterns of sharp Bragg reflection spots, and are ideal for analyzing the structure of solids.

TECHNIQUE

Some materials studied using crystallography, proteins for example, do not occur naturally as crystals. Typically, such molecules are placed in solution and allowed to crystallize over days, weeks, or months through vapor diffusion. A drop of solution containing the molecule, buffer, and precipitants is sealed in a container with a reservoir containing a hygroscopic solution. Water in the drop diffuses to the reservoir, slowly increasing the concentration and allowing a crystal to form. If the concentration were to rise more quickly, the molecule would simply precipitate out of solution, resulting in disorderly granules rather than an orderly and hence usable crystal.

Once a crystal is obtained, data can be collected using a beam of radiation. Although many universities that engage in crystallographic research have their own X-ray producing equipment, synchrotrons are often used as X-ray sources, because of the purer and more complete patterns such sources can generate. Synchrotron sources also have a much higher intensity of X-ray beams, so data collection takes a fraction of the time normally necessary at weaker sources.

Producing an image from a diffraction pattern requires sophisticated mathematics and often an iterative process of modelling and refinement. In this process, the mathematically predicted diffraction patterns of an hypothesized or "model" structure are compared to the actual pattern generated by the crystalline sample. Ideally, researchers make several initial guesses, which through refinement all converge on the same answer. Models are refined until their predicted patterns match to as great a degree as can be achieved without radical revision of the model. This is a painstaking process, made much easier today by computers.

The mathematical methods for the analysis of diffraction data only apply to *patterns,* which in turn result only when waves diffract from orderly arrays. Hence crystallography applies for the most part only to crystals, or to molecules which can be coaxed to crystallize for the sake of measurement. In spite of this, a certain amount of molecular information can be deduced from the patterns that are generated by fibers and powders, which while not as perfect as a solid crystal, may exhibit a degree of order. This level of order can be sufficient to deduce the structure of simple molecules, or to determine the coarse features of more complicated molecules (the double-helical structure of DNA, for example, was deduced from an X-ray diffraction pattern that had been generated by a fibrous sample).

CRYSTALLOGRAPHY IN MATERIALS ENGINEERING

Crystallography is a tool that is often employed by materials scientists. In single crystals, the effects of the crystalline arrangement of atoms is often easy to see macroscopically, because the natural shapes of crystals reflect the atomic structure. In addition, physical properties are often controlled by crystalline defects. The understanding of crystal structures is an important prerequisite for understanding crystallographic defects. Mostly, materials do not occur in a single crystalline, but poly-crystalline form, such that the powder diffraction method plays a most important role in structural determination.

A number of other physical properties are linked to crystallography. For example, the minerals in clay form small, flat, platelike structures. Clay can be easily deformed because the platelike particles can slip along each other in the plane of the plates, yet remain strongly connected in the direction perpendicular to the plates. Such mechanisms can be studied by crystallographic texture measurements.

In another example, iron transforms from a body-centered cubic (bcc) structure to a face-centered cubic (fcc) structure called austenite when it is heated. The fcc structure is a close-packed structure, and the bcc structure is not, which explains why the volume of the iron decreases when this transformation occurs.

Crystallography is useful in phase identification. When performing any process on a material, it may be desired to find out what compounds and what phases are present in the material. Each phase has a characteristic arrangement of atoms. Techniques like X-ray diffraction can be used to identify which patterns are present in the material, and thus which compounds are present (note: the determination of the "phases" within a material should not be confused with the more general problem of "phase determination," which refers to the phase of waves as they diffract from planes within a crystal, and which is a necessary step in the interpretation of complicated diffraction patterns).

Crystallography covers the enumeration of the symmetry patterns which can be formed by atoms in a crystal and for this reason has a relation to group theory and geometry. See symmetry group.

BIOLOGY

X-ray crystallography is the primary method for determining the molecular conformations of biological macromolecules, particularly protein and nucleic acids such as DNA and RNA. In fact, the double-helical structure of DNA was deduced from crystallographic data. The first crystal structure of a macromolecule was solved in 1958 (Kendrew, J.C. et al. (1958) A three-dimensional model of the myoglobin molecule obtained by X-ray analysis (*Nature* 181, 662–666). The Protein Data Bank (PDB) is a freely accessible repository for the structures of proteins and other biological macromolecules. Computer programs like RasMol or Pymol can be used to visualize biological molecular structures.

Electron crystallography has been used to determine some protein structures, most notably membrane proteins and viral capsids.

X-RAY CRYSTALLOGRAPHY

X-ray crystallography is a method of determining the arrangement of atoms within a crystal, in which a beam of X-rays strikes a crystal and scatters into many different directions. From the angles and intensities of these scattered beams, a crystallographer can produce a three-dimensional picture of the density of electrons within the crystal. From this electron density, the mean positions of the atoms in the crystal can be determined, as well as their chemical bonds, their disorder and various other information.

Since very many materials can form crystals — such as salts, metals, minerals, semiconductors, as well as various inorganic, organic and biological molecules — X-ray crystallography has been fundamental in the development of many scientific fields. In its first decades of use, this method determined the size of atoms, the lengths and types of chemical bonds, and the atomic-scale differences among various materials, especially minerals and alloys. The method also revealed the structure and functioning of many biological molecules, including vitamins, drugs, proteins and nucleic acids such as DNA. X-ray crystallography is still the chief method for characterizing the atomic structure of new materials and in discerning materials that appear similar by other experiments. X-ray crystal structures can also account for unusual electronic or elastic properties of a material, shed light on chemical interactions and processes, or serve as the basis for designing pharmaceuticals against diseases.

After a crystal has been obtained or grown in the laboratory, it is mounted on a goniometer and gradually rotated while being bombarded with X-rays, producing a diffraction pattern of regularly spaced spots known as *reflections*. The two-dimensional images taken at different rotations are converted into a three-dimensional model of the density of electrons within the crystal using the mathematical method of Fourier transforms, combined with chemical data known for the sample. Poor resolution (fuzziness) or even errors may result if the crystals are too small, or not uniform enough in their internal makeup.

Fig. 13.1: X-ray crystallography can locate every atom in a zeolite, an aluminosilicate with many important applications, such as water purification

X-ray crystallography is related to several other methods for determining atomic structures. Similar diffraction patterns can be produced by scattering electrons or neutrons, which are likewise interpreted as a Fourier transform. If single crystals of sufficient size cannot be obtained, various X-ray scattering methods can be applied to obtain less detailed information; such methods include fiber diffraction, powder diffraction and small-angle X-ray scattering (SAXS). In all these methods, the scattering is elastic; the scattered X-rays have the same wavelength as the incoming

X-ray. By contrast, *inelastic* X-ray scattering methods are useful in studying excitations of the sample, rather than the distribution of its atoms.

Early Scientific History of Crystals and X-rays

Crystals have long been admired for their regularity and symmetry, but they were not investigated scientifically until the 17th century. Johannes Kepler hypothesized in his work Strena seu de Nive Sexangula (1611) that the hexagonal symmetry of snowflake crystals was due to a regular packing of spherical water particles.

Crystals have long been admired for their regularity and symmetry, but they were not investigated scientifically until the 17th century. Johannes Kepler hypothesized in his work Strena seu de Nive Sexangula (1611) that the hexagonal symmetry of snowflake crystals was due to a regular packing of spherical water particles

X-rays were discovered by Wilhelm Conrad Röntgen in 1895, just as the studies of crystal symmetry were being concluded. Physicists were initially uncertain of the nature of X-rays, although it was soon suspected (correctly) that they were waves of electromagnetic radiation, in other words, another form of light. At that time, the wave model of light—specifically, the Maxwell theory of electromagnetic radiation—was well accepted among scientists, and experiments by Charles Glover Barkla showed that X-rays exhibited phenomena associated with electromagnetic waves, including transverse polarization and spectral lines akin to those observed in the visible wavelengths. Single-slit experiments in the laboratory of Arnold Sommerfeld suggested the wavelength of X-rays was roughly 1 Angström, one ten millionth of a millimetre. However, X-rays are composed of photons, and thus are not only waves of electromagnetic radiation but also exhibit particle-like properties. The photon concept was introduced by Albert Einstein in 1905 but it was not broadly accepted until 1922 when Arthur Compton confirmed it by the scattering of X-rays from electrons Therefore, these particle-like properties of X-rays, such as their ionization of gases, caused William Henry Bragg to argue in 1907 that X-rays were not electromagnetic radiation. Nevertheless, Bragg's view was not broadly accepted and the observation of X-ray diffraction in 1912 confirmed for most scientists that X-rays were a form of electromagnetic radiation.

X-rays Analysis of Crystals

Crystals are regular arrays of atoms, and X-rays can be considered waves of electromagnetic radiation. Atoms scatter X-ray waves, primarily through the atoms' electrons. Just as an ocean wave striking a lighthouse produces secondary circular waves emanating from the lighthouse, so an X-ray striking an electron produces secondary spherical waves emanating from the electron. This phenomenon is known as elastic scattering, and the electron (or lighthouse) is known as the scatterer. A regular array of scatterers produces a regular array of spherical waves. Although these waves cancel one another out in most directions through destructive interference, they add constructively in a few specific directions, determined by Bragg's law:

$$2d \sin\theta = n\lambda$$

where d is the spacing between diffracting planes, λ is the incident angle, n is any integer, and ? is the wavelength of the beam. These specific directions appear as spots on the diffraction pattern, often called *reflections*. Thus, X-ray diffraction results from an electromagnetic wave (the X-ray) impinging on a regular array of scatterers (the repeating arrangement of atoms within the crystal).

X-rays are used to produce the diffraction pattern because their wavelength ? is typically the same order of magnitude (1-100 Ångströms) as the spacing d between planes in the crystal. In principle, any wave impinging on a regular array of scatterers produces diffraction, as predicted first by Francesco Maria Grimaldi in 1665. To produce significant diffraction, the spacing between the scatterers and the wavelength of the impinging wave should be roughly similar in size. For illustration, the diffraction of sunlight through a bird's feather was first reported by James Gregory in the later 17th century. The first man-made diffraction gratings for visible light were constructed by David Rittenhouse in 1787, and Joseph von Fraunhofer in 1821. However, visible light has too long a wavelength (typically, 5500 Ångströms) to observe diffraction from crystals. However, prior to the first X-ray diffraction experiments, the spacings between lattice planes in a crystal were not known with certainty.

The idea that crystals could be used as a diffraction grating for X-rays arose in 1912 in a conversation between Paul Peter Ewald and Max von Laue in the English Garden in Munich. Ewald had proposed a resonator model of crystals for his thesis, but this model could not be validated using visible light, since the wavelength was much larger than the spacing between the resonators. Von Laue realized that electromagnetic radiation of a shorter wavelength was needed to observe such small spacings, and suggested that X-rays might have a wavelength comparable to the unit-cell spacing in crystals. Von Laue worked with two technicians, Walter Friedrich and his assistant Paul Knipping, to shine a beam of X-rays through a copper sulphate crystal and record its diffraction on a photographic plate. After being developed, the plate showed a large number of well-defined spots arranged in a pattern of intersecting circles around the spot produced by the central beam Von Laue developed a law that connects the scattering angles and the size and orientation of the unit-cell spacings in the crystal, for which he was awarded the Nobel Prize in Physics in 1914.

As described in the mathematical derivation below, the X-ray scattering is determined by the density of electrons within the crystal. Since the energy of an X-ray is much greater than that of an atomic electron, the scattering may be modeled as Thomson scattering, the interaction of an electromagnetic ray with a free electron. This model is generally adopted to describe the polarization of the scattered radiation. The intensity of Thomson scattering declines as $1/m^2$ with the mass m of the charged particle that is scattering the radiation; hence, the atomic nuclei, which are thousands of times heavier than an electron, contribute negligibly to the scattered X-rays.

Development from 1912 to 1920

After Von Laue's pioneering research, the field developed rapidly, most notably by physicists William Lawrence Bragg and his father William Henry Bragg. In 1912-1913, the younger Bragg developed Bragg's law, which connects the observed scattering with reflections from evenly spaced planes within the crystal. The earliest structures were generally simple and marked by one-dimensional symmetry. However, as computational and experimental methods improved over the next decades, it became feasible to deduce reliable atomic positions for more complicated two- and three-dimensional arrangements of atoms in the unit-cell.

The potential of X-ray crystallography for determining the structure of molecules and minerals—then only known vaguely from chemical and hydrodynamic experiments—was realized immediately. The earliest structures were simple inorganic crystals and minerals, but even these revealed fundamental laws of physics and chemistry. The first atomic-resolution structure to be solved (in 1914) was that of table salt (When an atomic structure is determined by X-ray crystallography, it is said to be "solved".) The distribution of electrons in the table-salt structure showed that crystals are not necessarily comprised of covalently bonded molecules, and proved the existence of ionic compounds. The structure of diamond was solved in the same year, proving the tetrahedral arrangement of its chemical bonds and showing that the C-C single bond was 1.52 Ångströms.

Other early structures included copper, calcium fluoride (CaF_2, also known as *fluorite*), calcite ($CaCO_3$) and pyrite (FeS_2) [in 1914; spinel ($MgAl_2O_4$) in 1915; the rutile and anatase forms of titanium dioxide (TiO_2) in 1916 pyrochroite and, by extension, brucite [$Mn(OH)_2$ and $Mg(OH)_2$, respectively] in 1919; and wurtzite (hexagonal ZnS) in 1920.

The structure of graphite was solved in 1916 by the related method of powder diffraction, which was developed by Peter Debye and Paul Scherrer and, independently, by Albert Hull in 1917 The structure of graphite was determined from single-crystal diffraction in 1924 by two groups independently Hull also used the powder method to determine the structures of various metals, such as iron and magnesium.

Contributions to Chemistry and Material Science

X-ray crystallography has led to a better understanding of chemical bonds and non-covalent interactions. The initial studies revealed the typical radii of atoms, and confirmed many theoretical models of chemical bonding, such as the tetrahedral bonding of carbon in the diamond structure, the octahedral bonding of metals observed in ammonium hexachloroplatinate (IV) and the resonance observed in the planar carbonate group and in aromatic molecules Kathleen Lonsdale's 1928 structure of hexamethylbenzene established the hexagonal symmetry of benzene and showed a clear difference in bond length between the aliphatic C-C bonds and aromatic C-C bonds; this finding led to the idea of resonance between chemical bonds, which had profound consequences for the development of chemistry. Her conclusions were anticipated by William Henry Bragg, who published models of naphthalene and anthracene in 1921 based on other molecules, an early form of molecular replacement.

Also in the 1920s, Victor Moritz Goldschmidt and later Linus Pauling developed rules for eliminating chemically unlikely structures and for determining the relative sizes of atoms. These rules led to the structure of brookite (1928) and an understanding of the relative stability of the rutile, brookite and anatase forms of titanium oxide.

The distance between two covalently bonded atoms is a sensitive measure of the bond strength and its bond order; thus, X-ray crystallographic studies have led to the discovery of even more exotic types of bonding in inorganic chemistry, such as metal-metal double bonds metal-metal quadruple bonds and three-center, two-electron bonds scattering experiment — has also provided evidence for the partially covalent character of hydrogen bonds. In the field of organometallic chemistry, the X-ray structure of ferrocene initiated scientific studies of sandwich compounds while that of Zeise's salt stimulated research into "back bonding" and metal-pi complexes in general. Finally, X-ray crystallography had a pioneering role in the development of supramolecular chemistry, particularly in clarifying the structures of the crown ethers and the principles of host-guest chemistry.

In material sciences, many complicated inorganic and organometallic systems have been analyzed using single-crystal methods, such as fullerenes, metalloporphyrins, and other complicated compounds. Single-crystal diffraction is also used in the pharmaceutical industry, due to recent problems with polymorphs. The major factors affecting the quality of single-crystal structures are the crystal's size and regularity; recrystallization is a commonly used technique to improve these factors in small-molecule crystals. The Cambridge Structural Database contains over 400,000 structures; over 99% of these structures were determined by X-ray diffraction.

Mineralogy and Metallurgy

Since the 1920s, X-ray diffraction has been the principal method for determining the arrangement of atoms in minerals and metals. The application of X-ray crystallography to mineralogy began with the structure of garnet, which was determined in 1924 by Menzer. A systematic X-ray crystallographic study of the silicates was undertaken in the 1920s. This study showed that, as the Si/O ratio is altered, the silicate crystals exhibit significant changes in their atomic arrangements. Machatschki extended these insights to minerals in which aluminium substitutes for the silicon atoms of the silicates. The first application of X-ray crystallography to metallurgy likewise occurred in the mid-1920s. Most notably, Linus Pauling's structure of the alloy Mg2Sn led to his theory of the stability and structure of complex ionic crystals.

Early Organic and Small Biological Molecules

The first structure of an organic compound, hexamethylenetetramine, was solved in 1923 This was followed by several studies of long-chain fatty acids, which are an important component of biological membranes. In the 1930s, the structures of much larger molecules with two-dimensional complexity began to be solved. A significant advance was the structure of phthalocyanine, a large planar molecule that is closely related to porphyrin molecules important in biology, such as heme, corrin and chlorophyll.

X-ray crystallography of biological molecules took off with Dorothy Crowfoot Hodgkin, who solved the structures of cholesterol (1937), vitamin B12 (1945) and penicillin (1954), for which she was awarded the Nobel Prize in Chemistry in 1964. In 1969, she succeeded in solving the structure of insulin, on which she worked for over thirty years.

Biological Macromolecular Crystallography

Crystal structures of proteins (which are irregular and hundreds of times larger than cholesterol) began to be solved in the late 1950s, beginning with the structure of sperm whale myoglobin by Max Perutz and Sir John Cowdery Kendrew, for which they were awarded the Nobel Prize in Chemistry in 1962. Since that success, over 39000 X-ray crystal structures of proteins, nucleic acids and other biological molecules have been determined. For comparison, the nearest competing method, nuclear magnetic resonance (NMR) spectroscopy has produced roughly 6000 structures. Moreover, crystallography can solve structures of arbitrarily large molecules, whereas solution-state NMR is restricted to relatively small molecules (less than 70 kDa). X-ray crystallography is now used routinely by scientists to determine how a pharmaceutical interacts with its protein target and what changes might be advisable to improve it. However, intrinsic membrane proteins remain challenging to crystallize because they require detergents or other means to solubilize them in isolation, and such detergents often interfere with crystallization. Such membrane proteins are a large component of the genome and include many proteins of great physiological importance, such as ion channels and receptors.

ELASTIC Vs. INELASTIC SCATTERING

X-ray crystallography is a form of elastic scattering; the outgoing X-rays have the same energy as the incoming X-rays, only with altered direction. Since the energy of a photon is inversely proportional to its wavelength, elastic scattering means that the outgoing photons have the same wavelength as the incoming photons. By contrast, *inelastic scattering* occurs when energy is transferred from the incoming X-ray to the crystal, e.g., by exciting an inner-shell electron to a higher energy level. Such inelastic scattering changes the wavelength of the outgoing beam, making it longer and less energetic. Inelastic scattering is useful for probing such excitations of matter, but are not as useful in determining the distribution of scatterers within the matter, which is the goal of X-ray crystallography.

X-rays range in wavelength from 10 to 0.01 nanometers; a typical wavelength used for crystallography is roughly 1 Å (0.1 nm), which is on the scale of covalent chemical bonds and the radius of a single atom. Longer-wavelength photons (such as ultraviolet radiation) would not have sufficient resolution to determine the atomic positions. At the other extreme, shorter-wavelength photons such as gamma rays are difficult to produce in large numbers, difficult to focus, and interact too strongly with matter, producing particle-antiparticle pairs. Therefore, X-rays are the "sweetspot" for wavelength when determining atomic-resolution structures from the scattering of electromagnetic radiation.

OTHER TYPES OF X-RAY SCATTERING

X-ray diffraction involves the scattering of X-rays from a single crystal. Other forms of elastic X-ray scattering include powder diffraction, SAXS and several types of X-ray fiber diffraction, which was used by Rosalind Franklin in determining the double-helix structure of DNA. In general, X-ray diffraction produces isolated spots ("reflections"), while the other methods produce smooth, continuous scattering. In general, X-ray diffraction offers more structural information than these other techniques; however, it requires a sufficiently large and regular crystal, which is not always possible to obtain.

All of these scattering methods generally use *monochromatic* X-rays, which are restricted to a single wavelength with minor deviations. A broad spectrum of X-rays (that is, a blend of X-rays with different wavelengths) can also be used to carry out X-ray diffraction, a technique known as the Laue method. This is the method used in the original discovery of X-ray diffraction. Laue scattering provides much structural information with only a short exposure to the X-ray beam, and is therefore used in structural studies of very rapid events (time-resolved X-ray crystallography). However, it is not as well-suited as monochromatic scattering for determining the full atomic structure of a crystal. It is better suited to crystals with relatively simple atomic arrangements, such as minerals.

The Laue back reflection mode records X-rays scattered backwards also from a broad spectrum source. This is useful if the sample is too thick or bulky for X-rays to transmit through it. The diffracting planes in the crystal are determined by knowing that the normal to the diffracting plane bisects the angle between the incident beam and the diffracted beam. A Greninger chart can be used to interpret the back reflection Laue photograph. The X-calibre RTXDB and MWL 110 are commercial systems for Laue back reflection pattern recording. This technique can be used in materials analysis or nondestructive testing.

ELECTRON AND NEUTRON DIFFRACTION

Other particles, such as electrons and neutrons, may be used to produce a diffraction pattern. Although electron, neutron, and X-ray scattering use very different equipment, the resulting diffraction patterns are analyzed using the same coherent diffraction imaging techniques.

As derived below, the electron density within the crystal and the diffraction patterns are related by a simple mathematical method, the Fourier transform, which allows the density to be calculated relatively easily from the patterns. However, this works only if the scattering is *weak*, i.e., if the scattered beams are much less intense than the incoming beam. Weakly scattered beams pass through the remainder of the crystal without undergoing a second scattering event. Such re-scattered waves are called "secondary scattering" and hinder the calculation of the density of scatterers. Any sufficiently thick crystal will produce secondary scattering but since X-rays interact relatively weakly with the electrons, this is generally not a significant concern. By contrast, electron beams may produce strong secondary scattering even for very small crystals (e.g., 100 μm) used in X-ray crystallography. In such cases, extremely thin samples, roughly 100 nanometers or less, must be used to avoid secondary scattering; the primary scattered electron beams leave the sample before they have a chance to undergo secondary scattering. Since this thickness corresponds roughly to the diameter of many viruses, a promising direction is the electron diffraction of isolated macromolecular assemblies, such as viral capsids and molecular machines, which may be carried out with a cryo-electron microscope.

Neutron diffraction is an excellent method for structure determination, although it has been difficult to obtain intense, monochromatic beams of neutrons in sufficient quantities. Traditionally, nuclear reactors have been used, although the new Spallation Neutron Source holds much promise in the near future. Being uncharged, neutrons scatter much more readily from the atomic nuclei rather than from the electrons. Therefore, neutron scattering is very useful for observing the positions of light atoms with few electrons, especially hydrogen, which is essentially invisible in the X-ray diffraction of larger molecules. Neutron scattering also has the remarkable property that the solvent can be made invisible by adjusting the ratio of normal water, H_2O, and heavy water, D_2O.

METHODS

Overview of Single-crystal X-ray Diffraction

The oldest and most precise method of X-ray crystallography is *single-crystal X-ray diffraction,* in which a beam of X-rays strikes a single crystal, producing scattered beams. When they land on a piece of film or other detector, these beams make a *diffraction pattern* of spots; the strengths and angles of these beams are recorded as the crystal is gradually rotated. Each spot is called a *reflection,* since it corresponds to the reflection of the X-rays from one set of evenly spaced planes within the crystal. For single crystals of sufficient purity and regularity, X-ray diffraction data can determine the mean chemical bond lengths and angles to within a few thousandths of an Ångström and to within a few tenths of a degree, respectively. The atoms in a crystal are also not static, but oscillate about their mean positions, usually by less than a few tenths of an Ångström. X-ray crystallography allows the size of these oscillations to be measured quantitatively.

Procedure

The technique of single-crystal X-ray crystallography has three basic steps. The first — and often most difficult — step is to obtain an adequate crystal of the material under study. The crystal should be sufficiently large (typically larger than 100 micrometres in all dimensions), pure in composition and regular in structure, with no significant internal imperfections such as cracks or twinning. A small or irregular crystal will give fewer and less reliable data, from which it may be impossible to determine the atomic arrangement.

In the second step, the crystal is placed in an intense beam of X-rays, usually of a single wavelength (*monochromatic X-rays*), producing the regular pattern of reflections. As the crystal is gradually rotated, previous reflections disappear and new ones appear; the intensity of every spot is recorded at every orientation of the crystal. Multiple data sets may have to be collected, with each set covering slightly more than half a full rotation of the crystal and typically containing tens of thousands of reflection intensities.

In the third step, these data are combined computationally with complementary chemical information to produce and refine a model of the arrangement of atoms within the crystal. The final, refined model of the atomic arrangement — now called a *crystal structure* — is usually stored in a public database.

Limitations

As the crystal's repeating unit, its unit cell, becomes larger and more complex, the atomic-level picture provided by X-ray crystallography becomes less well-resolved (more "fuzzy") for a given number of observed reflections. Two limiting cases of X-ray crystallography—"small-molecule" and "macromolecular" crystallography—are often discerned. *Small-molecule crystallography* typically involves crystals with fewer than 100 atoms in their asymmetric unit; such crystal structures are usually so well resolved that the atoms can be discerned as isolated "blobs" of electron density. By contrast, *macromolecular crystallography* often involves tens of thousands of atoms in the unit cell. Such crystal structures are generally less well-resolved (more "smeared out"); the atoms and chemical bonds appear as tubes of electron density, rather than as isolated atoms. In general, small molecules are also easier to crystallize than macromolecules; however, X-ray crystallography has proven possible even for viruses with hundreds of thousands of atoms.

Crystallization

Although crystallography can be used to characterize the disorder in an impure or irregular crystal, crystallography generally requires a pure crystal of high regularity to solve for the structure of a complicated arrangement of atoms. Pure, regular crystals can sometimes be obtained from natural or man-made materials, such as samples of metals, minerals or other macroscopic materials. The regularity of such crystals can sometimes be improved with annealing and other methods. However, in many cases, obtaining a diffraction-quality crystal is the chief barrier to solving its atomic-resolution structure.

Small-molecule and macromolecular crystallography differ in the range of possible techniques used to produce diffraction-quality crystals. Small molecules generally have few degrees of conformational freedom, and may be crystallized by a wide range of methods, such as chemical vapor deposition and recrystallisation. By contrast, macromolecules generally have many degrees of freedom and their crystallization must be carried out to maintain a stable structure. For example, proteins and larger RNA molecules cannot be crystallized if their tertiary structure has been unfolded; therefore, the range of crystallization conditions is restricted to solution conditions in which such molecules remain folded.

Protein crystals are almost always grown in solution. The most common approach is to lower the solubility of its component molecules very gradually; however, if this is done too quickly, the molecules will precipitate from solution, forming a useless dust or amorphous gel on the bottom of the container. Crystal growth in solution is characterized by two steps: *nucleation* of a microscopic crystallite (possibly having only 100 molecules), followed by *growth* of that crystallite, ideally to a diffraction-quality crystal The solution conditions that favor the first step (nucleation) are not always the same conditions that favor the second step (its subsequent growth). The crystallographer's goal is to identify solution conditions that favor the development of a single, large crystal, since larger crystals offer improved resolution of the molecule. Consequently, the solution conditions should *disfavor* the first step (nucleation) but *favor* the second (growth), so that only one large crystal forms per droplet. If nucleation is favored too much, a shower of small crystallites will form in the droplet, rather than one large crystal; if favored too little, no crystal will form whatsoever.

It is extremely difficult to predict good conditions for nucleation or growth of well-ordered crystals. In practice, favorable conditions are identified by *screening*; a very large batch of the molecules is prepared, and a wide variety of crystallization solutions are tested. Hundreds, even thousands, of solution conditions are generally tried before finding one that succeeds in crystallizing the molecules. The various conditions can use one or more physical mechanisms to lower the solubility of the molecule; for example, some may change the pH, some contain salts of the Hofmeister series or chemicals that lower the dielectric constant of the solution, and still others contain large polymers such as polyethylene glycol that drive the molecule out of solution by entropic effects. It is also common to try several temperatures for encouraging crystallization, or to gradually lower the temperature so that the solution becomes supersaturated. These methods require large amounts of the target molecule, as they use high concentration of the molecule(s) to be crystallized. Due to the difficulty in obtaining such large quantities (milligrams) of crystallisation grade protein, dispensing robots have been developed that are capable of accurately dispensing crystallisation trial drops that are of the order on 100 nanoliters in volume. This means that roughly 10-fold less protein is used per-experiment when compared to crystallisation trials setup by hand (on the order on 1 microliter).

Several factors are known to inhibit or mar crystallization. The growing crystals are generally held at a constant temperature and protected from shocks or vibrations that might disturb their crystallization. Impurities in the molecules or in the crystallization solutions are often inimical to crystallization. Conformational flexibility in the molecule also tends to make crystallization less likely, due to entropy. Ironically, molecules that tend to self-assemble into regular helices are often unwilling to assemble into crystals. Crystals can be marred by twinning, which can occur when a unit cell can pack equally favorably in multiple orientations; although recent advances in computational methods have begun to allow the structures of twinned crystals to be solved, it is still very difficult. Having failed to crystallize a target molecule, a crystallographer may try again with a slightly modified version of the molecule; even small changes in molecular properties can lead to large differences in crystallization behaviour.

DATA COLLECTION

Mounting the Crystal

Once they are full-grown, the crystals are mounted so that they may be held in the X-ray beam and rotated. There are several methods of mounting. Although crystals were once loaded into glass capillaries with the crystallization solution (the mother liquor), a more modern approach is to scoop the crystal up in a tiny loop, made of nylon or plastic and attached to a solid rod, that is then flash-frozen with liquid nitrogen. This freezing reduces the radiation damage of the X-rays, as well as the noise in the Bragg peaks due to thermal motion (the Debye-Waller effect). However, untreated crystals often crack if flash-frozen; therefore, they are generally pre-soaked in a cryoprotectant solution before freezing Unfortunately, this pre-soak may itself cause the crystal to crack, ruining it for crystallography. Generally, successful cryo-conditions are identified by trial and error.

The capillary or loop is mounted on a goniometer, which allows it to be positioned accurately within the X-ray beam and rotated. Since both the crystal and the beam are often very small, the crystal must be centered within the beam to within roughly 25 micrometres accuracy, which is aided by a camera focused on the crystal. The most common type of goniometer is the "kappa goniometer", which offers three angles of rotation: the φ angle, which rotates about an axis roughly perpendicular to the beam; the ω angle, about an axis at roughly 50° to the ω axis; and, finally, the φ angle about the loop/capillary axis. When the angle is zero, the ω and φ axes are aligned. The κ rotation allows for convenient mounting of the crystal, since the arm in which the crystal is mounted may be swung out towards the crystallographer. The oscillations carried out during data collection involve the ω axis only. An older type of goniometer is the four-circle goniometer, and its relatives such as the six-circle goniometer.

Recording the Reflections

When a crystal is mounted and exposed to an intense beam of X-rays, it scatters the X-rays into a pattern of spots or *reflections* that can be observed on a screen behind the crystal. A similar pattern may be seen by shining a laser pointer at a compact disc. The relative intensities of these spots provide the information to determine the arrangement of molecules within the crystal in atomic detail. The intensities of these reflections may be recorded with photographic film, an area detector or with a charge-coupled device (CCD) image sensor. The peaks at small angles correspond to low-resolution data, whereas those at high angles represent high-resolution data; thus, an upper limit on the eventual resolution of the structure can be determined from the first few images. Some measures of diffraction quality can be determined at this point, such as the mosaicity of the crystal and its overall disorder, as observed in the peak widths. Some pathologies of the crystal that would render it unfit for solving the structure can also be diagnosed quickly at this point.

One image of spots is insufficient to reconstruct the whole crystal; it represents only a small slice of the full Fourier transform. To collect all the necessary information, the crystal must be rotated step-by-step through 180°, with an image recorded at every step; actually, slightly more than 180° is required to cover reciprocal space, due to the curvature of the Ewald sphere. However, if the crystal has a higher symmetry, a smaller angle such as 90° or 45° may be recorded. The axis of the rotation should generally be changed at least once, to avoid developing a "blind spot" in reciprocal space close to the rotation axis. It is customary to rock the crystal slightly (by 0.5-2°) to catch a broader region of reciprocal space.

Multiple data sets may be necessary for certain phasing methods. For example, MAD phasing requires that the scattering be recorded at least three (and usually four, for redundancy) wavelengths of the incoming X-ray radiation. A single crystal may degrade too much during the collection of one data set, owing to radiation damage; in such cases, data sets on multiple crystals must be taken.

DATA ANALYSIS

Crystal Symmetry, Unit Cell, and Image Scaling

Having recorded a series of diffraction patterns from the crystal, each corresponding to a different crystal orientation, the crystallographer must now convert these two-dimensional images into a three-dimensional model of the density of electrons throughout the crystal using the mathematical technique of Fourier transforms. (The relevance of this technique is explained below.) Roughly speaking, each spot corresponds to a different type of variation in the electron density; the crystallographer must determine *which* variation corresponds to *which* spot (*indexing*), the relative strengths of the spots in different images (*merging and scaling*) and how the variations should be combined to yield the total electron density (*phasing*).

In order to process the data, a crystallographer must first *index* the reflections within the multiple images recorded. This means identifying the dimensions of the unit cell and which image peak corresponds to which position in reciprocal space. A byproduct of indexing is to determine the symmetry of the crystal, i.e., its *space group*. Some space groups can be eliminated from the beginning, since they require symmetries known to be absent in the molecule itself. For example, reflection symmetries cannot be observed in chiral molecules; thus, only 65 space groups of 243 possible are allowed for protein molecules which are almost always chiral. Indexing is generally accomplished using an *autoindexing* routine. Having assigned symmetry, the data is then *integrated*. This converts the hundreds of images containing the thousands of reflections into a single file, consisting of (at the very least) records of the Miller index of each reflection, and an intensity for each reflection [at this state the file often also includes error estimates and measures of partiality (what part of a given reflection was recorded on that image)].

A full data set may consist of hundreds of separate images taken at different orientations of the crystal. The first step is to merge and scale these various images, that is, to identify which peaks appear in two or more images (*merging*) and to scale the relative images so that they have a consistent intensity scale. Optimizing the intensity scale is critical because the relative intensity of the peaks is the key information from which the structure is determined. The repetitive technique of crystallographic data collection and the often high symmetry of crystalline materials cause the diffractometer to record many symmetry-equivalent reflections multiple times. This allows a merging or symmetry related R-factor to be calculated based upon how similar are the measured intensities of symmetry equivalent reflections, thus giving a score to assess the quality of the data.

Initial Phasing

The data collected from a diffraction experiment is a reciprocal space representation of the crystal lattice. The position of each diffraction 'spot' is governed by the size and shape of the unit cell, and the inherent symmetry within the crystal. The intensity of each diffraction 'spot' is recorded, and this intensity is proportional to the square of the *structure factor* amplitude. The structure factor is a complex number containing information relating to both the amplitude and phase of a wave. In order to obtain an interpretable *electron density map,* both amplitude and phase must be known (an electron density map allows a crystallographer to build a starting model of the molecule). The phase cannot be directly recorded during a diffraction experiment: this is known as the phase problem. Initial phase estimates can be obtained in a variety of ways:

- *Ab initio phasing,* or *direct methods:* This is usually the method of choice for small molecules (<1000 non-hydrogen atoms), and has been used successfully to solve the phase problems for small proteins. If the resolution of the data is better than 1.4 Å (140 pm), direct methods can be used to obtain phase information, by exploiting known phase relationships between certain groups of reflections.

- *Molecular replacement* - if a structure exists of a related structure, it can be used as a search model in molecular replacement to determine the orientation and position of the molecules within the unit cell. The phases obtained this way can be used to generate *electron density maps.*
- *Anomalous X-ray scattering* (*MAD or SAD phasing*) - the X-ray wavelength may be scanned past an absorption edge of an atom, which changes the scattering in a known way. By recording full sets of reflections at three different wavelengths one can solve for the substructure of the anomalously diffracting atoms and thence the structure of the whole molecule. The most popular method of incorporating anomalous scattering atoms into proteins is to express the protein in a methionine auxotroph (a host incapable of synthesising methionine) in a media rich in Seleno-methionine, which contains Selenium atoms. A MAD experiment can then be conducted around the absorption edge, which should then yield the position of any methionine residues within the protein, providing initial phases.
- *Heavy atom methods* (*i.e MIR*) - If electron-dense metal atoms can be introduced into the crystal, direct methods or Patterson-space methods can be used to determine their location and to obtain initial phases. Typically, a crystallographer can introduce such heavy atoms either by soaking the crystal in a heavy atom-containing solution, or by co-crystallization (growing the crystals in the presence of a heavy atom). As in MAD phasing, the changes in the scattering amplitudes can be

interpreted to yield the phases. Although this is the original method by which protein crystal structures were solved, it has largely been superseded by MAD phasing with selenomethionine.

While all four of the above methods are used to solve the phase problem for protein crystallography, small molecule crystallography generally yields data suitable for structure solution using direct methods (*ab initio* phasing).

Model Building and Phase Refinement

Having obtained initial phases, an initial model can be built. This model can be used to refine the phases, leading to an improved model, and so on. Given a model of some atomic positions, these positions and their respective Debye-Waller factors (accounting for the thermal motion of the atom - aka B-factors) can be refined to fit the observed diffraction data, ideally yielding a better set of phases. A new model can then be fit to the new electron density map and a further round of reffnement is carried out. This continues until the correlation between the diffraction data and the model is maximized. The agreement is measured by an *R*-factor defined as

$$R = \frac{\sum_{\text{all reflections}} |F_0 - F_c|}{\sum_{\text{all reflections}} |F_0|}$$

A similar quality criterion is R_{free}, which is calculated from a subset (~10%) of reflections that were not included in the structure refinement. Both *R* factors depend on the resolution of the data. As a rule of thumb, R_{free} should be approximately the resolution in Ångströms divided by 10; thus, a data-set with 2 Å resolution should yield a final R_{free} of roughly 0.2. Chemical bonding features such as stereochemistry, hydrogen bonding and distribution of bond lengths and angles are complementary measures of the model quality. Phase bias is a serious problem in such iterative model building. *Omit maps* are a common technique used to check for this.

It may not be possible to observe every atom of the crystallized molecule - it must be remembered that the resulting electron density is an average of all the molecules within the crystal. In some cases, there is too much residual disorder in those atoms, and the resulting electron density for atoms existing in many conformations is smeared to such an extent that it is no longer detectable in the electron density map. Weakly scattering atoms such as hydrogen are routinely invisible. It is also possible for a single atom to appear multiple times in an electron density map, e.g., if a protein sidechain has multiple (<4) allowed conformations. In still other cases, the crystallographer may detect that the covalent structure deduced for the molecule was incorrect, or changed. For example, proteins may be cleaved or undergo post-translational modifications that were not detected prior to the crystallization.

DEPOSITION OF THE STRUCTURE

Once the model of a molecule's structure has been finalized, it is often deposited in a crystallographic database such as the Protein Data Bank (for protein structures) or the Cambridge Structural Database (for small molecules). Many structures obtained in private commercial ventures to crystallize medicinally relevant proteins, are not deposited in public crystallographic databases.

DIFFRACTION THEORY

Further information: Dynamical theory of diffraction and Bragg diffraction.

The main goal of X-ray crystallography is to determine the density of electrons $f(\mathbf{r})$ throughout the crystal, where **r** represents the three-dimensional position vector within the crystal. To do this, X-ray scattering is used to collect data about its Fourier transform $F(q)$, which is inverted mathematically to obtain the density defined in real space, using the formula

$$F(r) = \int \frac{dq}{(2\pi)^3} F(\mathbf{q}) e^{i\mathbf{q}.\mathbf{r}}$$

where the integral is summed over all possible values of q. The three-dimensional real vector **q** represents a point in reciprocal space, that is, to a particular oscillation in the electron density as one moves in the direction in which **q** points. The length of **q** corresponds to 2π divided by the wavelength of the oscillation. The corresponding formula for a Fourier transform will be used below:

$$F(\mathbf{q}) = \int dr f(r) e^{-i\mathbf{q}.r}$$

where the integral is summed over all possible values of the position vector r within the crystal.

The Fourier transform $F(q)$ is generally a complex number, and therefore has a magnitude $|F(q)|$ and a phase $f(q)$ related by the equation

$$F(\mathbf{q}) = |F(\mathbf{q})| e^{i\phi(\mathbf{q})}$$

The intensities of the reflections observed in X-ray diffraction give us the magnitudes $|F(q)|$ but not the phases $f(q)$. To obtain the phases, full sets of reflections are collected with known alterations to the scattering, either by modulating the wavelength past a certain absorption edge or by adding strongly scattering (i.e., electron-dense) metal atoms such as mercury. Combining the magnitudes and phases yields the full Fourier transform $F(\mathbf{q})$, which may be inverted to obtain the electron density $f(\mathbf{r})$.

Crystals are often idealized as being *perfectly* periodic. In that ideal case, the atoms are positioned on a perfect lattice, the electron density is perfectly periodic, and the Fourier transform $F(\mathbf{q})$ is zero except when **q** belongs to the reciprocal lattice (the so-called *Bragg peaks*). In reality, however, crystals are not perfectly periodic; atoms vibrate about their mean position, and there may be disorder of various types, such as mosaicity, dislocations, various point defects, and heterogeneity in the conformation of crystallized molecules. Therefore, the Bragg peaks have a finite width and there may be significant *diffuse scattering*, a continuum of scattered X-rays that fall between the Bragg peaks.

Intuitive Understanding by Bragg's Law

An intuitive understanding of X-ray diffraction can be obtained from the Bragg model of diffraction. In this model, a given reflection is associated with a set of evenly spaced sheets running through the crystal, usually passing through the centers of the atoms of the crystal lattice. The orientaticn of a particular set of sheets is identified by its three Miller indices (*h*, *k*, *l*), and let their spacing be noted by *d*. William Lawrence Bragg proposed a model in which the incoming X-rays are scattered specularly (mirror-like) from each plane; from that assumption, X-rays scattered from adjacent planes will combine constructively (constructive interference) when the angle λ between the plane and the X-ray results in a path-length difference that is an integer multiple *n* of the X-ray wavelength λ.

$$2d \sin \theta = n\lambda$$

A reflection is said to be *indexed* when its Miller indices (or, more correctly, its reciprocal lattice vector components) have been identified from the known wavelength and the scattering angle 2θ. Such indexing gives the unit-cell parameters, the lengths and angles of the unit-cell, as well as its space group. Since Bragg's law does not interpret the relative intensities of the reflections, however, it is generally inadequate to solve for the arrangement of atoms within the unit-cell; for that, a Fourier transform method must be carried out.

Scattering as a Fourier Transform

The incoming X-ray beam has a polarization and should be represented as a vector wave; however, for simplicity, let it be represented here as a scalar wave. We also ignore the complication of the time dependence of the wave and just focus on the wave's spatial dependence. Plane waves can be represented by a wave vector k_{in}, and so the strength of the incoming wave at time $t=0$ is given by

$$Ae^{ikin.r}$$

At position **r** within the sample, let there be a density of scatterers f(**r**); these scatterers should produce a scattered spherical wave of amplitude proportional to the local amplitude of the incoming wave times the number of scatterers in a small volume dV about **r**

amplitude of scattered wave = $Ae^{ik.r}\ S\,f(\mathbf{r})\ dV$

where S is the proportionality constant.

Let s consider the fraction of scattered waves that leave with an outgoing wave-vector of $\mathbf{k}_{out}$ and strike the screen at $\mathbf{r}_{screen}$. Since no energy is lost (elastic, not inelastic scattering), the wavelengths are the same as are the magnitudes of the wave-vectors $|\mathbf{k}_{in}| = |\mathbf{k}_{out}|$. From the time that the photon is scattered at r until it is absorbed at $\mathbf{r}_{screen}$, the photon undergoes a change in phase

$$e^{i\mathbf{k}_{out}.(\mathbf{r}_{screen}-\mathbf{r})}$$

The net radiation arriving at rscreen is the sum of all the scattered waves throughout the crystal

$$AS\int d\mathbf{r}\ f(\mathbf{r})e^{i\mathbf{k}in.\mathbf{r}}e^{i\mathbf{k}_{out}.(\mathbf{r}_{screen}-\mathbf{r})}$$

$$= ASe^{i\mathbf{k}_{out}.\mathbf{r}_{screen}}\int d\mathbf{r}\ f(\mathbf{r})e^{i(\mathbf{k}_{in}-\mathbf{k}_{out}).\mathbf{r}}$$

which may be written as a Fourier transform

$$ASe^{i\mathbf{k}_{out}.\mathbf{r}_{screen}}\int d\mathbf{r}\ f(\mathbf{r})e^{-i\mathbf{q}.\mathbf{r}} = ASe^{i\mathbf{k}_{out}.\mathbf{r}_{screen}}F(\mathbf{q})$$

where $q = k_{out} - k_{in}$. The measured intensity of the reflection will be square of this amplitude

$$A^2 S^2 \left|F(\boldsymbol{q})\right|^2$$

Advantages of a Crystal

In principle, an atomic structure could be determined from applying X-ray scattering to non-crystalline samples, even to a single molecule. However, crystals offer a much stronger signal due to their periodicity.

A crystalline sample is by definition periodic; a crystal is composed of many unit cells repeated indefinitely in three independent directions. Such periodic systems have a Fourier transform that is concentrated at periodically repeating points in reciprocal space known as Bragg peaks; the Bragg peaks correspond to the reflection spots observed in the diffraction image. Since the amplitude at these reflections grows linearly with the number N of scatterers, the observed intensity of these spots should grow quadratically, like N^2. In other words, using a crystal concentrates the weak scattering of the individual unit cells into a much more powerful, coherent reflection that can be observed above the noise. This is an example of constructive interference.

In a non-crystalline sample, molecules within that sample would be in random orientations and therefore would have a continuous Fourier spectrum that spreads its amplitude more uniformly and with a much reduced intensity, as is observed in SAXS. More importantly, the orientational information is lost. In the crystal, the molecules adopt the same orientation within the crystal, whereas in a liquid, powder or amorphous state, the observed signal is averaged over the possible orientations of the molecules. Although theoretically possible with sufficiently low-noise data, it is generally difficult to obtain atomic-resolution structures of complicated, asymmetric molecules from such rotationally averaged scattering data. An intermediate case is fiber diffraction in which the subunits are arranged periodically in at least one dimension.

CHAPTER - 14

Cyclic Voltammetry

Cyclic voltammetry or CV is a type of potentiodynamic electrochemical measurement. In a cyclic voltammetry experiment the working electrode potential is ramped linearly versus time like linear sweep voltammetry. Cyclic voltammetry takes the experiment a step farther than linear sweep voltammetry which ends when it reaches a set

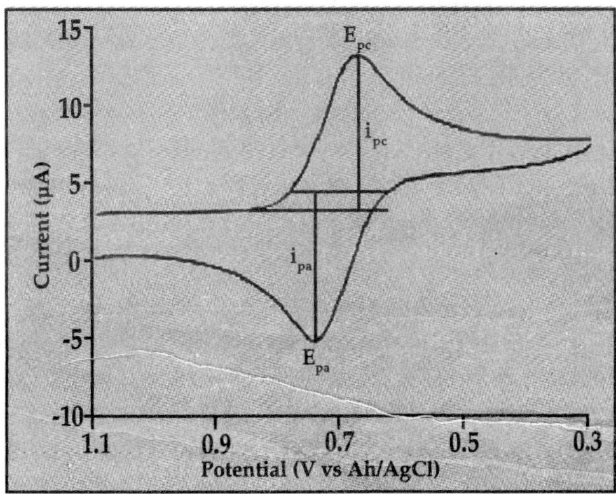

Fig. 14.1: Typical cyclic voltammogram

potential. When cyclic voltammetry reaches a set potential, the working electrode's potential ramp is inverted. This inversion can happen multiple times during a single experiment. The current at the working electrode is plotted versus the applied voltage to give the cyclic voltammogram trace. Cyclic voltammetry is generally used to study the electrochemical properties of an analyte in solution.

EXPERIMENTAL METHOD

In cyclic voltammetry, the electrode potential ramps linearly versus time as shown. This ramping is known as the experiment's scan rate (V/s). The potential is measured between the reference electrode and the working electrode and the current is measured between the working electrode and the counter electrode. This data is then plotted as current (i) vs. potential (E). As the waveform shows, the forward scan produces a current peak for any analytes that can be reduced (or oxidized depending on the initial scan direction) through the range of the potential scanned. The current will increase as the potential reaches the reduction potential of the analyte, but then falls off as the concentration of the analyte is depleted close to the electrode surface. If the redox couple is reversible then when the applied potential is reversed, it will reach the potential that will reoxidize the product formed in the first reduction reaction, and produce a current of reverse polarity from the forward scan. This oxidation peak will usually have a similar shape to the reduction peak. As a result, information about the redox potential and electrochemical reaction rates of the compounds are obtained.

For instance if the electronic transfer at the surface is fast and the current is limited by the diffusion of species to the electrode surface, then the current peak will be proportional to the square root of the scan rate. This relationship is described by the Cottrell equation.

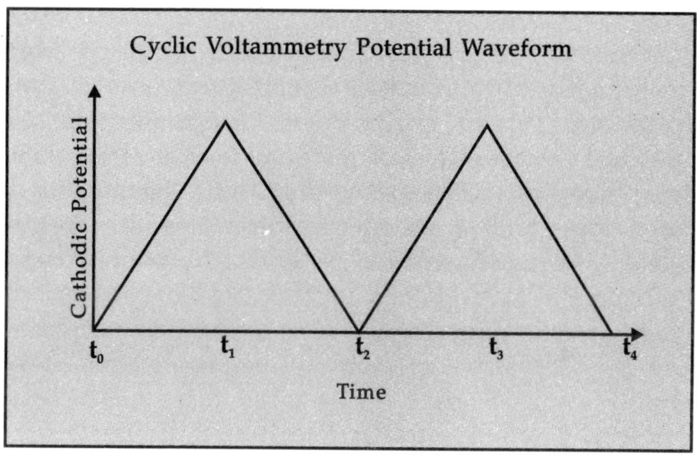

Fig. 14.2: Cyclic voltammetry waveform

CHARACTERIZATION

The utility of cyclic voltammetry is highly dependent on the analyte being studied. The analyte has to be redox active within the experimental potential window. It is also highly desirable for the analyte to display a reversible wave. A reversible wave is when an analyte is reduced or oxidized on a forward scan and is then reoxidized or rereduced in a predictable way on the return scan as shown in the first figure.

Reversible peaks have a distinct absolute potential difference between the reduction (E_{pc}) and oxidation peak (E_{pa}). In an ideal system $|E_{pc}-E_{pa}|$ would be 59 mV for a 1 electron process and 30 mV for a 2 electron process. In addition the ratio of the currents passed at reduction (i_{pc}) and end oxidation (i_{pa}) is near unity ($1 = i_{pa}/i_{pc}$) for a reversible peaks.

When such reversible peaks are observed thermodynamic information in the form of half cell potential $E^0_{½}$ can be determined. When waves are semi-reversible such as when i_{pa}/i_{pc} is less than or greater than 1, it can be possible to determine even more information especially kinetic processes like following chemical reaction.

When waves are non-reversible it is impossible to determine what their thermodynamic $E^0_{½}$ is with cyclic voltammetry. This $E^0_{½}$ can be determined however it often requires equal quantities of the analyte in both oxidation states. When a wave is non-reversible cyclic voltammetry can not determine if the wave is at its thermodynamic potential or shift to a more extreme potential by some form of overpotential. The couple could be irreversible because of a following chemical process, a common example for transition metals is a shift in the geometry of the coordination sphere. If this is the case, then higher scan rates may show a reversible wave. It is also possible that the wave is irreversible due to a physical process most commonly some form of precipitation as discussed below. Some speculation can be made in regards to irreversible waves however they are generally outside the scope of cyclic voltammetry.

EXPERIMENTAL SETUP

The method uses a reference electrode, working electrode, and counter electrode which in combination are sometimes referred to as a three-electrode setup. Electrolyte is usually added to the test solution to ensure sufficient conductivity. The combination of the solvent, electrolyte and specific working electrode material determines the range of the potential.

Electrodes are static and sit in unstirred solutions during cyclic voltammetry. This "still" solution method results in cyclic voltammetry's characteristic diffusion controlled peaks. This method also allows a portion of the analyte to remain after reduction or oxidation where it may display further redox activity. Stirring the solution between cyclic voltammetry traces is important as to supply the electrode surface with fresh analyte for each new experiment. The solubility of an analyte can change drastically with its overall charge. Since cyclic voltammetry usually alters the charge of the analyte it is common for reduced or oxidized analyte to precipitate out onto the electrode. This layering of analyte can insulate the electrode surface, display its own redox activity in subsequent scans, or at the very least alter the electrode surface. For this and other reasons it is often necessary to clean electrodes between scans.

Common materials for working electrodes include glassy carbon, platinum, and gold. These electrodes are generally encased in a rod of inert insulator with a disk exposed at one end. A regular working electrode has a radius within an order of magnitude of 1 mm. Having a controlled surface area with a defined shape is important for interpreting cyclic voltammetry results.

To run cyclic voltammetry experiments at high scan rates a regular working electrode is insufficient. High scan rates create peaks with large currents and increased resistances which result in distortions. Ultramicroelectrodes can be used to minimize the current and resistance.

The counter electrode, also known as the auxiliary or second electrode, can be any material which conducts easily and won't react with the bulk solution. Reactions occurring at the counter electrode surface are unimportant as long as it continues to conduct current well. To maintain the observed current the counter electrode will often oxidize or reduce the solvent or bulk electrolyte.

Reference electrodes are a complex subject and worth investigating elsewhere.

VARIATIONS

In some experiments an electroactive species is fixed to the surface of the electrode, for instance in microparticle voltammetry.

Potentiodynamic techniques also exist that add low-amplitude ac perturbation to a potential ramp and measure variable response in a single frequency (ac voltammetry) or in many frequencies simultaneously (potentiodynamic electrochemical impedance spectroscopy). The response in alternating current is two-dimensional – it is characterised by amplitude and phase. The amplitude and phase depend differently on frequency for constituents of ac response attributed to different processes (charge transfer, diffusion, double layer charging, etc.). Frequency response analysis enables simultaneous monitoring of the various processes that contribute to the potentiodynamic ac response of electrochemical system.

DISTINCTIONS

Cyclic voltammetry is not a hydrodynamic technique. In a hydrodynamic technique flow is achieved at the electrode surface by stirring the solution, pumping the solution, or rotating the electrode as is the case with rotating disk electrodes and rotating ring-disk electrodes. These techniques target steady state conditions which appear the same scanned from the positive or the negative, thus limiting them to linear sweep voltammetry.

MASS SPECTROMETRY

Mass spectrometry (MS) is an analytical technique for the determination of the elemental composition of a sample or molecule. It is also used for elucidating the chemical structures of molecules, such as peptides and other chemical compounds. The MS principle consists of ionizing chemical compounds to generate charged molecules or molecule fragments and measurement of their mass-to-charge ratios In a typical MS procedure, a sample is loaded onto the MS instrument, and its compounds are ionized by different methods (e.g., by impacting them with an electron beam), resulting in the formation of charged particles (ions). The mass-to-charge ratio of the particles is then calculated from the motion of the ions as they transit through electromagnetic fields.

MS instruments consist of three modules: an *ion source,* which splits the sample molecules into ions; a *mass analyzer,* which sorts the ions by their masses by applying electromagnetic fields; and a *detector,* which measures the value of an indicator quantity and thus provides data for calculating the abundances of each ion present. The technique has both qualitative and quantitative uses. These include identifying unknown compounds, determining the isotopic composition of elements in a molecule, and determining the structure of a compound by observing its fragmentation. Other uses include quantifying the amount of a compound in a sample or studying the fundamentals of gas phase ion chemistry (the chemistry of ions and neutrals in a vacuum). MS is now in very common use in analytical laboratories that study physical, chemical, or biological properties of a great variety of compounds.

ETYMOLOGY

The word *spectrograph* has been used since 1884 as an "*International Scientific Vocabulary*". The linguistic roots are a combination and removal of bound morphemes and free morphemes which relate to the terms ***spectr**-um* and *phot-**ograph-ic** plate* Early *spectrometry* devices that measured the mass-to-charge ratio of ions were called *mass spectrographs* which consisted of instruments that recorded a spectrum of mass values on a photographic plate A *mass spectroscope* is similar to a *mass spectrograph* except that the beam of ions is directed onto a phosphor screen A mass spectroscope configuration was used in early instruments when it was desired that the effects of adjustments be quickly observed. Once the instrument was properly adjusted, a photographic plate was inserted and exposed. The term mass spectroscope continued to be used even though the direct illumination of a phosphor screen was replaced by indirect measurements with an oscilloscope. The use of the term *mass spectroscopy* is now discouraged due to the possibility of confusion with light spectroscopy Mass spectrometry is often abbreviated as *mass-spec* or simply as *MS* Thomson has also noted that a *mass spectroscope* is similar to a *mass spectrograph* except that the beam of ions is directed onto a phosphor screen The suffix -scope here denotes the direct viewing of the spectra (range) of masses.

HISTORY

In 1886, Eugen Goldstein observed rays in gas discharges under low pressure that traveled away from the anode and through channels in a perforated cathode, opposite to the direction of negatively charged cathode rays (which travel from cathode to anode). Goldstein called these positively charged anode rays "Kanalstrahlen"; the standard translation of this term into English is "canal rays". Wilhelm Wien found that strong electric or magnetic fields deflected the canal rays and, in 1899, constructed a device with parallel electric and magnetic fields that separated the positive rays according to their charge-to-mass ratio (Q/m). Wien found that the charge-to-mass ratio depended on the nature of the gas in the discharge tube. English scientist J.J. Thomson later improved on the work of Wien by reducing the pressure to create a mass spectrograph.

Some of the modern techniques of mass spectrometry were devised by Arthur Jeffrey Dempster and F.W. Aston in 1918 and 1919 respectively. In 1989, half of the Nobel Prize in Physics was awarded to Hans Dehmelt and Wolfgang Paul for the development of the ion trap technique in the 1950s and 1960s. In 2002, the Nobel Prize in Chemistry was awarded to John Bennett Fenn for the development of electrospray ionization (ESI) and Koichi Tanaka for the development of soft laser desorption (SLD) in 1987. However earlier, matrix-assisted laser desorption/ionization (MALDI), was developed by Franz Hillenkamp and Michael Karas; this technique has been widely used for protein analysis.

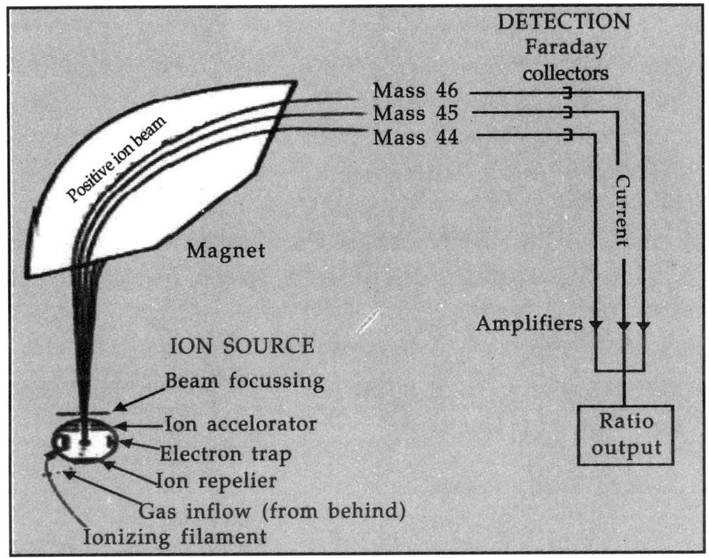

Fig. 14.3: Schematics of a simple mass spectrometer with sector type mass analyzer

SIMPLIFIED EXAMPLE

The following example describes the operation of a spectrometer mass analyzer, which is of the sector type. (Other analyzer types are treated below.) Consider a sample of sodium chloride (table salt). In the ion source, the sample is vaporized (turned into gas) and ionized (transformed into electrically charged particles) into sodium (Na^+) and chloride (Cl^-) ions. Sodium atoms and ions are monoisotopic, with a mass of about 23 amu. Chloride atoms and ions come in two isotopes with masses of approximately 35 amu (at a natural abundance of about 75%) and approximately 37 amu (at a natural abundance of about 25%). The analyzer part of the spectrometer contains electric and magnetic fields, which exert forces on ions traveling through these fields. The speed of a charged particle may be increased or decreased while passing through the electric field, and its direction may be altered by the magnetic field. The magnitude of the deflection of the moving ion's trajectory depends on its mass-to-charge ratio. By Newton's second law of motion, lighter ions get deflected by the magnetic force more than heavier ions. The streams of sorted ions pass from the analyzer to the detector, which records the relative abundance of each ion type. This information is used to determine the chemical element composition of the original sample (i.e. that both sodium and chlorine are present in the sample) and the isotopic composition of its constituents (the ratio of ^{35}Cl to ^{37}Cl).

INSTRUMENTATION

Ion Source Technologies

The ion source is the part of the mass spectrometer that ionizes the material under analysis (the analyte). The ions are then transported by magnetic or electric fields to the mass analyzer.

Techniques for ionization have been key to determining what types of samples can be analyzed by mass spectrometry. Electron ionization and chemical ionization are used for gases and vapors. In chemical ionization sources, the analyte is ionized by chemical ion-molecule reactions during collisions in the source. Two techniques often used with liquid and solid biological samples include electrospray ionization (invented by John Fenn) and matrix-assisted laser desorption/ionization (MALDI, developed by K. Tanaka and separately by M. Karas and F. Hillenkamp).

Inductively coupled plasma (*ICP*) sources are used primarily for cation analysis of a wide array of sample types. In this type of Ion Source Technology, a 'flame' of plasma that is electrically neutral overall, but that has had a substantial fraction of its atoms ionised by high temperature) is used to atomize introduced sample molecules and to further strip the outer electrons from those atoms. The plasma is usually generated from argon gas, since the first ionization energy of argon atoms is higher than the first of any other elements except He, O, F and Ne, but lower than the second ionization energy of all except the most electropositive metals. The heating is achieved by a radio-frequency current passed through a coil surrounding the plasma. Others include glow discharge, field desorption (FD), fast atom bombardment (FAB), thermospray, desorption/ionization on silicon (DIOS), Direct Analysis in Real Time (DART), atmospheric pressure chemical ionization (APCI), secondary ion mass spectrometry (SIMS), spark ionization and thermal ionization (TIMS).

Ion Attachment Ionization is a newer soft ionization technique that allows for fragmentation free analysis.

A FT-ICR MASS SPECTROMETER

Fourier Transform Mass Spectrometry

Fourier transform mass spectrometry, or more precisely Fourier transform ion cyclotron resonance MS, measures mass by detecting the image current produced by ions cyclotroning in the presence of a magnetic field. Instead of measuring the deflection of ions with a detector such as an electron multiplier, the ions are injected into a Penning trap (a static electric/magnetic ion trap) where they effectively form part of a circuit. Detectors at fixed positions in space measure the electrical signal of ions which pass near them over time, producing a periodic signal. Since the frequency of an ion's cycling is determined by its mass to charge ratio, this can be deconvoluted by performing a Fourier transform on the signal. FTMS has the advantage of high sensitivity (since each ion is "counted" more than once) and much higher resolution and thus precision.

Ion cyclotron resonance (ICR) is an older mass analysis technique similar to FTMS except that ions are detected with a traditional detector. Ions trapped in a Penning trap are excited by an RF electric field until they impact the wall of the trap, where the detector is located. Ions of different mass are resolved according to impact time.

Orbitraps

Very similar nonmagnetic FTMS has been performed, where ions are electrostatically trapped in an orbit around a central, spindle shaped electrode. The electrode confines the ions so that they both orbit around the central electrode and oscillate back and forth along the central electrode's long axis. This oscillation generates an image current in the detector plates which is recorded by the instrument. The frequencies of these image currents depend on the mass to charge ratios of the ions. Mass spectra are obtained by Fourier transformation of the recorded image currents.

Similar to Fourier transform ion cyclotron resonance mass spectrometers, Orbitraps have a high mass accuracy, high sensitivity and a good dynamic range.

Detector

The final element of the mass spectrometer is the detector. The detector records either the charge induced or the current produced when an ion passes by or hits a surface. In a scanning instrument, the signal produced in the detector during the course of the scan versus where the instrument is in the scan (at what m/Q) will produce a mass spectrum, a record of ions as a function of m/Q.

Typically, some type of electron multiplier is used, though other detectors including Faraday cups and ion-to-photon detectors are also used. Because the number of ions leaving the mass analyzer at a particular instant is typically quite small, considerable amplification is often necessary to get a signal. Microchannel Plate Detectors are commonly used in modern commercial instruments In FTMS and Orbitraps, the detector consists of a pair of metal surfaces within the mass analyzer/ion trap region which the ions only pass near as they oscillate. No DC current is produced, only a weak AC image current is produced in a circuit between the electrodes. Other inductive detectors have also been used.

TANDEM MASS SPECTROMETRY

A tandem mass spectrometer is one capable of multiple rounds of mass spectrometry, usually separated by some form of molecule fragmentation. For example, one mass analyzer can isolate one peptide from many entering a mass spectrometer. A second mass analyzer then stabilizes the peptide ions while they collide with a gas, causing them to fragment by collision-induced dissociation (CID). A third mass analyzer then sorts the fragments produced from the peptides. Tandem MS can also be done in a single mass analyzer over time, as in a quadrupole ion trap. There are various methods for fragmenting molecules for tandem MS, including collision-induced dissociation (CID), electron capture dissociation (ECD), electron transfer dissociation (ETD), infrared multiphoton dissociation (IRMPD) and blackbody infrared radiative dissociation (BIRD). An important application using tandem mass spectrometry is in protein identification.

Tandem mass spectrometry enables a variety of experimental sequences. Many commercial mass spectrometers are designed to expedite the execution of such routine sequences as single reaction monitoring (SRM), multiple reaction monitoring (MRM), and precursor ion scan. In SRM, the first analyzer allows only a single mass through and the second analyzer monitors for a single user defined fragment ion. MRM allows for multiple user defined fragment ions. SRM and MRM are most often used with scanning instruments where the second mass analysis event is duty cycle limited. These experiments are used to increase specificity of detection of known molecules, notably in pharmacokinetic studies. Precursor ion scan refers to monitoring for a specific loss from the precursor ion. The first and second mass analyzers scan across the spectrum as partitioned by a user defined *m/z* value. This experiment is used to detect specific motifs within unknown molecules.

An important type of Tandem mass spectrometry is Accelerator Mass Spectrometry (AMS), which uses very high voltages, usually in the mega-volt range, to accelerate negative ions into a type of tandem mass spectrometer. One of the most important applications of this technique is radiocarbon dating.

COMMON MASS SPECTROMETER CONFIGURATIONS AND TECHNIQUES

When a specific configuration of source, analyzer, and detector becomes conventional in practice, often a compound acronym arises to designate it, and the compound acronym may be more well known among nonspectrometrists than the component acronyms. The epitome of this is MALDI-TOF, which simply refers to combining a matrix-assisted laser desorption/ionization source with a time-of-flight mass analyzer. The MALDI-TOF moniker is more widely recognized by the non-mass spectrometrist scientist than MALDI or TOF individually. Other examples include inductively coupled plasma-mass spectrometry (ICP-MS), accelerator mass spectrometry (AMS), Thermal ionization-mass spectrometry (TIMS) and spark source mass spectrometry (SSMS). Sometimes the use of the generic "MS" actually connotes a very specific mass analyzer and detection system, as is the case with AMS, which is always sector based.

Certain applications of mass spectrometry have developed monikers that although strictly speaking they would seem to refer to a broad application, in practice have come instead to connote a specific or a limited number of instrument configurations. An example of this is isotope ratio mass spectrometry (IRMS), which refers in practice to the use of a limited number of sector based mass analyzers; this name is used to refer to both the application and the instrument used for the application.

CHROMATOGRAPHIC TECHNIQUES COMBINED WITH MASS SPECTROMETRY

An important enhancement to the mass resolving and mass determining capabilities of mass spectrometry is using it in tandem with chromatographic separation techniques.

Gas Chromatography

A common combination is gas chromatography mass spectrometry (GC/MS or GC-MS). In this technique, a gas chromato-graph is used to separate different compounds. This stream of separated compounds is fed online into the ion source, a metallic filament to which voltage is applied. This filament emits electrons which ionize the compounds. The ions can then further fragment, yielding predictable patterns. Intact ions and fragments pass into the mass spectrometer's analyzer and are eventually detected.

Liquid Chromatography

Similar to gas chromatography MS (GC/MS), liquid chromatography mass spectrometry (LC/MS or LC-MS) separates compounds chromatographically before they are introduced to the ion source and mass spectrometer. It differs from GC/MS in that the mobile phase is liquid, usually a mixture of water and organic solvents, instead of gas. Most commonly, an electrospray ionization source is used in LC/MS. There are also some newly developed ionization techniques like laser spray.

Ion Mobility

Ion mobility spectrometry/mass spectrometry (IMS/MS or IMMS) is a technique where ions are first separated by drift time through some neutral gas under an applied electrical potential gradient before being introduced into a mass spectrometer. Drift time is a measure of the radius relative to the charge of the ion. The duty cycle of IMS (the time over which the experiment takes place) is longer than most mass spectrometric techniques, such that the mass spectrometer can sample along the course of the IMS separation. This produces data about the IMS separation and the mass-to-charge ratio of the ions in a manner similar to LC/MS.

The duty cycle of IMS is short relative to liquid chromatography or gas chromatography separations and can thus be coupled to such techniques, producing triple modalities such as LC/IMS/MS.

DATA AND ANALYSIS

Mass spectrometry produces various types of data. The most common data representation is the mass spectrum.

Certain types of mass spectrometry data are best represented as a mass chromatogram. Types of chromatograms include selected ion monitoring (SIM), total ion current (TIC), and selected reaction monitoring chromatogram (SRM), among many others.

Other types of mass spectrometry data are well represented as a three dimensional contour map. In this form, the mass-to-charge, m/z is on the x-axis, intensity the y-axis, and an additional experimental parameter, such as time, is recorded on the z-axis.

Data Analysis

Basics

Mass spectrometry data analysis is a complicated subject matter that is very specific to the type of experiment producing the data. There are general subdivisions of data that are fundamental to understand any data.

Many mass spectrometers work in either *negative ion mode* or *positive ion mode*. It is very important to know whether the observed ions are negatively or positively charged. This is often important in determining the neutral mass but it also indicates something about the nature of the molecules.

Different types of ion source result in different arrays of fragments produced from the original molecules. An electron ionization source produces many fragments and mostly odd electron species with one charge, whereas an electrospray source usually produces quasimolecular even electron species that may be multiply charged. Tandem mass spectrometry purposely produces fragment ions post-source and can drastically change the sort of data achieved by an experiment.

By understanding the origin of a sample, certain expectations can be assumed as to the component molecules of the sample and their fragmentations. A sample from a synthesis/manufacturing process will likely contain impurities chemically related to the target component. A relatively crudely prepared biological sample will likely contain a certain amount of salt, which may form adducts with the analyte molecules in certain analyses.

Results can also depend heavily on how the sample was prepared and how it was run/introduced. An important example is the issue of which matrix is used for MALDI spotting, since much of the energetics of the desorption/ionization event is controlled by the matrix rather than the laser power. Sometimes samples are spiked with sodium or another ion-carrying species to produce adducts rather than a protonated species.

The greatest source of trouble when non-mass spectrometrists try to conduct mass spectrometry on their own or collaborate with a mass spectrometrist is inadequate definition of the research goal of the experiment. Adequate definition of the experimental goal is a prerequisite for collecting the proper data and successfully interpreting it. Among the determinations that can be achieved with mass spectrometry are molecular mass, molecular structure, and sample purity. Each of these questions requires a different experimental procedure. Simply asking for a "mass spec" will most likely not answer the real question at hand.

Interpretation of Mass Spectra

Since the precise structure or peptide sequence of a molecule is deciphered through the set of fragment masses, the interpretation of mass spectra requires combined use of various techniques. Usually the first strategy for identifying an unknown compound is to compare its experimental mass spectrum against a library of mass spectra. If the search comes up empty, then manual interpretation or software assisted interpretation of mass spectra are performed. Computer simulation of ionization and fragmentation processes occurring in mass spectrometer is the primary tool for assigning structure or peptide sequence to a molecule. An *a priori* structural information is fragmented *in silico* and the resulting pattern is compared with observed spectrum. Such simulation is often supported by a fragmentation library hat contains published patterns of known decomposition reactions. Software taking advantage of this idea has been developed for both small molecules and proteins.

Another way of interpreting mass spectra involves spectra with accurate mass. A mass-to-charge ratio value (*m/z*) with only integer precision can represent an immense number of theoretically possible ion structures. More precise mass figures significantly reduce the number of candidate molecular formulas, albeit each can still represent large number of structurally diverse compounds. A computer algorithm called formula generator calculates all molecular formulas that theoretically fit a given mass with specified tolerance.

A recent technique for structure elucidation in mass spectrometry, called precursor ion fingerprinting identifies individual pieces of structural information by conducting a search of the tandem spectra of the molecule under investigation against a library of the product-ion spectra of structurally characterized precursor ions.

APPLICATIONS

Isotope Ratio MS: Isotope Dating and Tracking

Mass spectrometry is also used to determine the isotopic composition of elements within a sample. Differences in mass among isotopes of an element are very small, and the less abundant isotopes of an element are typically very rare, so a very sensitive instrument is required. These instruments, sometimes referred to as isotope ratio mass spectrometers (IR-MS), usually use a single magnet to bend a beam of ionized particles towards a series of Faraday cups which convert particle impacts to electric current. A fast on-line analysis of deuterium content of water can be done using Flowing afterglow mass spectrometry, FA-MS. Probably the most sensitive and accurate mass spectrometer for this purpose is the accelerator mass spectrometer (AMS). Isotope ratios are important markers of a variety of processes. Some isotope ratios are used to determine the age of materials for example as in carbon dating. Labelling with stable isotopes is also used for protein quantification.

Trace Gas Analysis

Several techniques use ions created in a dedicated ion source injected into a flow tube or a drift tube: selected ion flow tube (SIFT-MS), and proton transfer reaction (PTR-MS), are variants of chemical ionization dedicated for trace gas analysis of air, breath or liquid headspace using well defined reaction time allowing calculations of analyte concentrations from the known reaction kinetics without the need for internal standard or calibration.

Atom Probe

An atom probe is an instrument that combines time-of-flight mass spectrometry and field ion microscopy (FIM) to map the location of individual atoms.

Pharmacokinetics

Pharmacokinetics is often studied using mass spectrometry because of the complex nature of the matrix (often blood or urine) and the need for high sensitivity to observe low dose and long time point data. The most common instrumentation used in this application is LC-MS with a triple quadrupole mass spectrometer. Tandem mass spectrometry is usually employed for added specificity. Standard curves and internal standards are used for quantitation of usually a single pharmaceutical in the samples. The samples represent different time points as a pharmaceutical is administered and then metabolized or cleared from the body. Blank or t=0 samples taken before administration are important in determining background and insuring data integrity with such complex sample matrices. Much attention is paid to the linearity of the standard curve; however it is not uncommon to use curve fitting with more complex functions such as quadratics since the response of most mass spectrometers is less than linear across large concentration ranges.

There is currently considerable interest in the use of very high sensitivity mass spectrometry for microdosing studies, which are seen as a promising alternative to animal experimentation.

Protein Characterization

Mass spectrometry is an important emerging method for the characterization of proteins. The two primary methods for ionization of whole proteins are electrospray ionization (ESI) and matrix-assisted laser desorption/ionization (MALDI). In keeping with the performance and mass range of available mass spectrometers, two approaches are used for characterizing proteins. In the first, intact proteins are ionized by either of the two techniques described above, and then introduced to a mass analyser.

This approach is referred to as "top-down" strategy of protein analysis. In the second, proteins are enzymatically digested into smaller peptides using proteases such as trypsin or pepsin, either in solution or in gel after electrophoretic separation. Other proteolytic agents are also used. The collection of peptide products are then introduced to the mass analyser. When the characteristic pattern of peptides is used for the identification of the protein the method is called peptide mass fingerprinting (PMF), if the identification is performed using the sequence data determined in tandem MS analysis it is called de novo sequencing. These procedures of protein analysis are also referred to as the "bottom-up" approach.

Space Exploration

As a standard method for analysis, mass spectrometers have reached other planets and moons. Two were taken to Mars by the Viking program. In early 2005 the Cassini-Huygens mission delivered a specialized GC-MS instrument aboard the Huygens probe through the atmosphere of Titan, the largest moon of the planet Saturn. This instrument analyzed atmospheric samples along its descent trajectory and was able to vaporize and analyze samples of Titan's frozen, hydrocarbon covered surface once the probe had landed. These measurements compare the abundance of isotope(s) of each particle comparatively to earth's natural abundance. Also onboard the Cassini-Huygens spacecraft is an ion and neutral mass spectrometer which has been taking measurements of Titan's atmospheric composition as well as the composition of Enceladus' plumes.

Mass spectrometers are also widely used in space missions to measure the composition of plasmas. For example, the Cassini spacecraft carries the Cassini Plasma Spectrometer (CAPS), which measures the mass of ions in Saturn's magnetosphere.

Respired Gas Monitor

Mass spectrometers were used in hospitals for respiratory gas analysis beginning around 1975 through the end of the century. Some are likely still in use but none are currently being manufactured.

Found mostly in the operating room, they were a part of a complex system in which respired gas samples from patients undergoing anesthesia were drawn into the instrument through a valve mechanism designed to sequentially connect up to 32 rooms to the mass spectrometer. A computer directed all operations of the system. The data collected from the mass spectrometer was delivered to the individual rooms for the anesthesiologist to use.

This magnetic sector mass spectrometer's uniqueness may have been the fact that a plane of detectors, each purposely positioned to collect all of the ion species expected to be in the samples, allowed the instrument to simultaneously report all of the patient respired gases. Although the mass range was limited to slightly over 120 u, fragmentation of some of the heavier molecules negated the need for a higher detection limit.

CHAPTER - 15

Transmission Electron Microscopy

Transmission electron microscopy (TEM) is a microscopy technique whereby a beam of electrons is transmitted through an ultra thin specimen, interacting with the specimen as they pass through. An image is formed from the interaction of the electrons transmitted through the specimen, which is magnified and focused onto an imaging device, such as a fluorescent screen, as is common in most TEMs, on a layer of photographic film, or to be detected by a sensor such as a CCD camera.

TEMs are capable of imaging at a significantly higher resolution than light microscopes, owing to the small de Broglie wavelength of electrons. This enables the instrument to be able to examine fine detail — even as small as a single column of atoms, which is tens of thousands times smaller than the smallest resolvable object in a light microscope. TEM forms a major analysis method in a range of scientific fields, in both physical and biological sciences. TEMs find application in cancer research, virology, materials science as well as pollution and semiconductor research.

At smaller magnifications TEM image contrast is due to absorption of electrons in the material, due to the thickness and composition of the material. At higher magnifications complex wave interactions modulate the intensity of the image, requiring expert analysis of observed images. Alternate modes of use allow for the TEM to observe modulations in chemical identity, crystal orientation, electronic structure and sample induced electron phase shift as well as the regular absorption based imaging.

The first TEM was built by Max Knoll and Ernst Ruska in 1931, with this group developing the first TEM with resolving power greater than that of light in 1933 and the first commercial TEM in 1939.

INITIAL DEVELOPMENT

Ernst Abbe originally proposed that the ability to resolve detail in an object was limited by the wavelength of the light used in imaging, thus limiting the useful obtainable magnification from an optical microscope to a few micrometres. Developments into UV microscopes, led by Koehler, allowed for an increase in resolving power of about a factor of two. However this required more expensive quartz optical components, due to the absorption of UV by glass. At this point it was believed that obtaining an image with sub-micrometre information was simply impossible due to this wavelength constraint.

It had earlier been recognized by Plücker in 1858 that the deflection of "cathode rays" (electrons) was possible by the use of magnetic fields. This effect had been utilised to build primitive cathode ray oscilloscopes (CROs) as early as 1897 by Ferdinand Braun, intended as a measurement device. Indeed in 1891 it was recognized by Riecke that the cathode rays could be focussed by these magnetic fields, allowing for simple lens designs. Later this theory was extended by Hans Busch in his work published in 1926, who showed that the lens maker's equation, could under appropriate assumptions, be applicable to electrons.

In 1928, at the Technological University of Berlin Adolf Matthias, Professor of High voltage Technology and Electrical Installations, appointed Max Knoll to lead a team of researchers to advance the CRO design. The team consisted of several PhD students including Ernst Ruska and Bodo von Borries. This team of researchers concerned themselves with lens design and CRO column placement, which they attempted to obtain the parameters that could be optimised to allow for construction of better CROs, as well as the development of electron optical components which could be used to generate low magnification (nearly 1:1) images. In 1931 the group successfully generated magnified images of mesh grids placed over the anode aperture. The device used two magnetic lenses to achieve higher magnifications, arguably the first electron microscope.

IMPROVING RESOLUTION

At this time the wave nature of electrons, which were considered charged matter particles, had not been fully realised until the publication of the De Broglie hypothesis in 1927 The group was unaware of this publication until 1932, where it was quickly realized that the De Broglie wavelength of electrons was many orders of magnitude smaller than that for light, theoretically allowing for imaging at atomic scales. In April 1932, Ruska suggested the construction of a new electron microscope for direct imaging of specimens inserted into the microscope, rather than simple mesh grids or images of apertures. With this device successful diffraction and normal imaging of aluminium sheet was achieved, however exceeding the magnification achievable with light microscopy had still not been successfully demonstrated. This goal was achieved in September 1933, using images of cotton fibers, which were quickly acquired before being damaged by the electron beam.

At this time, interest in the electron microscope had increased, with other groups, such as Albert Prebus and James Hillier at the University of Toronto who constructed the first TEM in north America in 1938, continually advancing TEM design.

Research continued on the electron microscope at Siemens in 1936, the aim of the research was the development improvement of TEM imaging properties, particularly with regard to biological specimens. In 1939 the first commercial electron microscope, pictured, was installed in the Physics department of I. G Farben-Werke. Further work on the electron microscope was hampered by the destruction of a new laboratory constructed at Siemens by an air-raid, as well as the death of two of the researchers, Heinz Müller and Friedrick Krause during World War II.

FURTHER RESEARCH

After world war II, Ruska resumed work at Siemens, where he continued to develop the electron microscope, producing the first 100kx microscope. The fundamental structure of this microscope design, with multi-stage beam preparation optics, is still used in modern microscopes.

With the development of TEM, the associated technique of scanning transmission electron microscopy (STEM) was re-investigated and did not become developed until the 1970s, with Albert Crewe at the University of Chicago developing the field emission gun[8] and adding a high quality objective lens to create the modern STEM. Using this design, Crewe demonstrated the ability to image atoms using annular dark-field imaging. Crewe and coworkers at the University of Chicago developed the cold field emission electron source and built a STEM able to visualize single heavy atoms on thin carbon substrates.

BACKGROUND

Electrons

Theoretically, the maximum resolution, d, that one can obtain with a light microscope has been limited by the wavelength of the photons that are being used to probe the sample, ? and the numerical aperture of the system, NA.

$$d = \frac{\lambda}{2n\sin\alpha} \approx \frac{\lambda}{2\,\mathrm{NA}}$$

Early twentieth century scientists theorised ways of getting around the limitations of the relatively large wavelength of visible light (wavelengths of 400–700 nanometers) by using electrons. Like all matter, electrons have both wave and particle properties (as theorized by Louis-Victor de Broglie), and their wave-like properties mean that a beam of electrons can be made to behave like a beam of electromagnetic radiation. The wavelength of electrons is found to be given by equating the de Broglie equation to the kinetic energy of an electron. An additional correction must be made to account for relativistic effects, as in a TEM an electron's velocity approaches the speed of light, c.

$$\lambda_e \approx \frac{h}{\sqrt{2m_0E(1+\frac{E}{2m_oc^2})}}$$

whose, h is Planck's constant, m_0 is the rest mass of an electron and E is the energy of the accelerated electron.

Electrons are usually generated in an electron microscope by a process known as thermionic emission from a filament, usually tungsten, in the same manner as a light bulb, or alternatively by field emission. The electrons are then accelerated by an electric potential (measured in volts) and focused by electrostatic and electromagnetic lenses onto the sample. The transmitted beam contains information about electron density, phase and periodicity; this beam is used to form an image.

Source Formation

From the top down, the TEM consists of an emission source, which may be a tungsten filament, or a lanthanum hexaboride (LaB_6) source. For tungsten, this will be of the form of either a hairpin-style filament, or a small spike-shaped filament. LaB_6 sources utilize small single crystals. By connecting this gun to an HV source (Typically ~120 kV for many applications) the gun will, given sufficient current, begin to emit electrons either by thermionic or field emission into the vacuum. This extraction is usually aided by the use of a Wehnelt cylinder. Once extracted, the upper lenses of the TEM allow for the formation of the electron probe to the desired size and location for later interaction with the sample.

Manipulation of the electron beam is performed using two physical effects. The interaction of electrons with a magnetic field will cause electrons to move according to the right hand rule, thus allowing for electromagnets to manipulate the electron beam. The use of magnetic fields allows for the formation of a magnetic lens of variable focusing power, the lens shape originating due to the distribution of magnetic flux. Additionally, electrostatic fields can cause the electrons to be deflected through a constant angle. Coupling of two deflections in opposing directions with a small intermediate gap allows for the formation of a shift in the beam path, this being used in TEM for beam shifting, subsequently this is extremely important to STEM. From these two effects, as well as the use of an electron imaging system, sufficient control over the beam path is possible for TEM operation. Additionally, the optical configuration of a TEM can be rapidly changed, unlike that for an optical microscope, as lenses in the beam path can be enabled, have their strength changed, or be disabled entirely simply via rapid electrical switching, the speed of which is only limited by the magnetic hysteresis of the lenses.

Optics

The lenses of a TEM allow for beam convergence, with the angle of convergence as a variable parameter, giving the

TEM the ability to change magnification simply by modifying the amount of current that flows through the coil, quadrupole or hexapole lenses. The quadrupole lens is an arrangement of electromagnetic coils at the vertices of the square, enabling the generation of a lensing magnetic fields, the hexapole configuration simply enhances the lens symmetry by using six, rather than four coils.

Typically a TEM consists of three stages of lensing. The stages are the condensor lenses, the objective lenses, and the projector lenses. The condensor lenses are responsible for primary beam formation, whilst the objective lenses focus the beam down onto the sample itself. The projector lenses are used to expand the beam onto the phosphor screen or other imaging device, such as film. The magnification of the TEM is due to the ratio of the distances between the specimen and the objective lens' image plane. Additional quad or hexpole lenses allow for the correction of asymmetrical beam distortions, known as astigmatism. It is noted that TEM optical configurations differ significantly with implementation, with manufacturers using custom lens configurations, such as in spherical aberration corrected instruments, or TEMs utilising energy filtering to correct electron chromatic aberration.

Display

Imaging systems in a TEM consist of a phosphor screen, which may be made of fine (10-100 um) particulate zinc sulphide, for direct observation by the operator. Optionally, an image recording system such as film based or doped YAG screen coupled CCDs Typically these devices can be removed or inserted into the beam path by the operator as required.

COMPONENTS

A TEM is composed of several components, which include a vacuum system in which the electrons travel, an electron emission source for generation of the electron stream, a series of electromagnetic lenses, as well as electrostatic plates. The latter two allow the operator to guide and manipulate the beam as required. Also required is a device to allow the insertion into, motion within, and removal of specimens from the beam path. Imaging devices are subsequently used to create an image from the electrons that exit the system.

Vacuum System

In order to allow for uninterrupted passage of electrons, the TEM must be evacuated to low pressures, typically on the order of 10^{-4} to 10^{-8} kPa The need for this is twofold, firstly the allowance for the voltage difference between the cathode and the ground without generating an arc, and secondly to reduce the collision frequency of electrons with gas atoms to negligible levels — this effect is characterised by the mean free path. As the TEM, unlike a CRT, is a system where components must be replaced, specimens inserted and – particularly on older TEMs – film cartridges must be replenished, the ability to re-evacuate a TEM on a regular basis is required. As such, TEMs are equipped with extensive pumping systems and are not permanently vacuum sealed.

To evacuate a TEM, the vacuum system consists of several stages. High pressure "low vacuum" pumps such as Diaphragm pumps are utilized to perform evacuation of the TEM to a sufficiently low pressure to allow the operation of turbomolecular or diffusion pumps which can reduce the TEM's in column pressure even further. To allow for the low vacuum pump to not require continuous operation, whilst accommo-dating continuous operation of the turbomolecular pumps, the vacuum side of a low pressure pump may be connected to evacuated chambers which serves no other purpose but the accommodation of the turbomolecular exhaust gases. TEMs may be isolated into chambers by the use of gate valves, which allow for sections of the TEM to be subjected to differing vacuum quality. This is often done as a preventative measure to prevent inadvertent loss of vacuum in critical locations, such as the electron gun in high resolution TEMs.

For high voltage TEMs, even very high vacuum conditions (very low pressure) are sufficient to allow for the generation of an electrical arc, particularly at the TEM cathode As such for higher voltage TEMs a third vacuum system may operate, with the gun isolated from the main chamber either by use of gate valves or by the use of a differential pumping aperture, which is essentially a small hole that prevents gas molecules from entering a high quality vacuum system faster than they can be pumped out. For these very low pressures either an ion pump or a getter material is used.

Poor vacuum in a TEM can cause several problems, from deposition of gas inside the TEM onto the specimen as it is being viewed through a process known as electron beam induced deposition, or in more severe cases damage to the cathode from a electrical discharge. Vacuum problems owing to specimen sublimation are limited by the use of a cold trap to adsorb sublimated gases in the vicinity of the specimen.

Specimen Stage

The specimen stage design allows for the external insertion of a TEM sample into the vacuum, ideally with minimal disruption to the vacuum inside the TEM. To allow for the transport of samples between multiple TEMs, a common standard of sample "grid" design is a 3 mm diameter brass ring, with a thickness of 100 um and an inner diameter of approximately 2.5 mm into which the sample is placed. The TEM stage and holder pair are designed to accommodate such specimens, although a wide variety of designs of stages and holder exist.

Once inserted into a TEM the sample often has to be manipulated to present the region of interest to the beam, sometimes — such as in single grain diffraction — in a specific orientation. To accommodate this, the TEM stage includes a mechanism for the translation (XYZ) and often rotation of the sample. Thus a TEM stage may provide four degrees of freedom for the motion of the specimen. Additional degrees of freedom can be provided by specialised holder designs. Of note however is that some stage designs, such as so-called "vertical insertion" stages, may simply only have X-Y translation available. The design criteria of TEM stages are complex, owing to the simultaneous requirements of mechanical and electron-optical constraints and have thus generated many unique implementations.

A TEM stage is required to have the ability to hold a specimen and direct the region of interest to the beam path. As the TEM can operate over a wide range of magnifications, the stage must simultaneously be highly resistant to mechanical drift, with drift requirements as low as a few nm/ minute whilst be able to be driven at speeds of from to several mm/minute down to several um/minute. Earlier designs of TEM accomplished this with a complex set of mechanical downgearing devices, allowing the operator to finely control the motion of the stage by several rotating rods. Modern devices may use electrical stage designs, using screw gearing in concert with stepper motors, providing the operator with a computer-based stage input, such as a joystick or trackball.

Two main designs for stages in a TEM exist, the top entry and the side-entry version. Each design must accommodate the matching holder to allow for specimen insertion without either damaging delicate TEM optics nor allowing gas into the TEM vacuum.

The top entry holder consists of a cartridge that is several cm long with a bore drilled down the cartridge axis. The specimen is loaded into the bore, possibly utilising a small screw ring to hold the sample in place. This cartridge is inserted into an airlock with the bore perpendicular to the TEM optic axis. When sealed, the airlock is manipulated to push the cartridge such that the cartridge falls into place, where the bore hole becomes aligned with the beam axis, such that the beam travels down the cartridge bore and into the specimen. Such designs are typically unable to be tilted without blocking the beam path or interfering with the objective lens.

The second design is the side entry holder, where the specimen is placed near the tip of a long metal (brass or stainless steel) rod, with the specimen placed flat in a small bore. Along the rod are several polymer vacuum rings to allow for the formation of a vacuum seal of sufficient quality, when inserted into the stage. The stage is thus designed to accommodate the rod, placing the sample either in between or near the objective lens, dependent upon the objective design. When inserted into the stage, the side entry holder has its tip contained within the TEM vacuum, and the base is presented to atmosphere, the airlock formed by the vacuum rings.

Insertion procedures for side entry TEM holders typically involve the rotation of the sample to trigger micro switches that initiate different pump cycles to minimise vacuum leakage into the chamber.

Contrast Formation

Contrast formation in the TEM depends greatly on the mode of operation. Complex imaging techniques which utilize the unique ability to change lens strength or to deactivate a lens allows for many operating modes. These modes may be used to discern information that is of particular interest to the investigator.

Bright Field

The most common mode of operation for a TEM is the bright field imaging mode. In this mode the contrast formation, when considered classically, is formed directly by occlusion and absorption of electrons in the sample. Thicker regions of the sample, or regions with a higher atomic number will appear dark, whilst regions with no sample in the beam path will appear bright – hence the term "bright field".

Diffraction Contrast

Samples can exhibit diffraction contrast, whereby the electron beam undergoes Bragg scattering, which in the case of a crystalline sample, disperses electrons into discrete locations in the back focal plane. By the placement of apertures in the back focal plane, i.e. the objective aperture, the desired Bragg reflections can be selected (or excluded), thus only parts of the sample that are causing the electrons to scatter to the selected reflections will end up projected onto the imaging apparatus.

If the reflections that are selected do not include the unscattered beam (which will appear up at the focal point of the lens), then the image will appear dark wherever no sample scattering to the selected peak is present, as such a region without a specimen will appear dark. This is known as a dark-field image.

Modern TEMs are often equipped with specimen holders that allow the user to tilt the specimen to a range of angles in order to obtain specific diffraction conditions, and apertures placed above the specimen allow the user to select electrons that would otherwise be diffracted in a particular direction from entering the specimen.

Applications for this method include the identification of lattice defects in crystals. By carefully selecting the orientation of the sample, it is possible not just to determine the position of defects but also to determine the type of defect present. If the sample is orientated so that one particular plane is only slightly tilted away from the strongest diffracting angle (known as the Bragg Angle), any distortion of the crystal plane that locally tilts the plane to the Bragg angle will produce particularly strong contrast variations.

However, defects that produce only displacement of atoms that do not tilt the crystal to the Bragg angle (i. e. displacements parallel to the crystal plane) will not produce strong contrast.

Electron Energy Loss

Utilizing the advanced technique of EELS, for TEMs appropriately equipped electrons can be rejected based upon their voltage (which, due to constant charge is their energy), using magnetic sector based devices known as EELS spectrometers. These devices allow for the selection of particular energy values, which can be associated with the way the electron has interacted with the sample. For example different elements in a sample result in different electron energies in the beam after the sample. This normally results in chromatic aberration – however this effect can, for example, be used to generate an image which provides information on elemental composition, based upon the atomic absorption of electrons during electron-electron interaction.

EELS spectrometers can often be operated in both a spectroscopic and imaging modes, allowing for isolation or rejection of elastically scattered beams. As for many images inelastic scattering will include information that may not be of interest to the investigator thus reducing observable signals of interest, EELS imaging can be used to enhance contrast in observed images, including both bright field and diffraction, by rejecting unwanted components.

Phase Contrast

Crystal structure can also be investigated by High Resolution Transmission Electron Microscopy (HRTEM), also known as phase contrast. When utilizing a Field emission source, the images are formed due to differences in phase of electron waves, which is caused by specimen interaction. Image formation is given by the complex modulus of the incoming electron beams. As such, the image is not only dependent on the number of electrons hitting the screen, making direct interpretation of phase contrast images more complex. However this effect can be used to advantage, as it can be manipulated provide more information about the sample, such as in complex phase retrieval techniques.

Diffraction

As previously stated, by adjusting the magnetic lenses such that the back focal plane of the lens is placed on the imaging apparatus a diffraction pattern can be generated.

For thin crystalline samples, this produces an image that consists of a series of dots in the case of a single crystal, or a series of rings in the case of a polycrystalline material. For the single crystal case the diffraction pattern is dependent upon the orientation of the specimen. This image provides the investigator with information about the space group symmetries in the crystal and the crystal's orientation to the beam path. This is typically done without utilizing any information but the position at which the diffraction spots appear and the observed image symmetries.

Diffraction patterns can have a large dynamic range, and for crystalline samples, may have intensities greater than that which can be recorded by CCD. As such, TEMs may still be equipped with film cartridges for the purpose of obtaining these images

Analysis of diffraction patterns beyond point-position can be complex, as the image is sensitive to a number of factors such as specimen thickness and orientation, objective lens defocus, spherical and chromatic aberration. Although quantitative interpretation of the contrast shown in lattice images is possible, it is inherently complicated

More complex diffraction behavior is also possible, with phenomena such as Kikuchi lines or convergent beam electron diffraction (CBED) providing additional information, beyond structural data, such as sample thickness.

Three Dimensional Imaging

As TEM specimen holders typically allow for the rotation of a sample by a desired angle, multiple views of the same specimen can be obtained by rotating the angle of the sample along an axis perpendicular to the beam. By taking multiple images of a single TEM sample at differing angles, typically in 1 degree increments, a set of images known as a "tilt series" can be collected. Under purely absorption contrast conditions, this set of images can be used to construct a three dimensional representation of the sample. The reconstruction is accomplished by a two step process, first images are aligned to account for errors in the positioning of a sample; such errors can occur due to vibration or mechanical drift. Alignment methods use image registration algorithms, such as autocorrelation methods to correct these errors. Secondly, using a technique known as filtered back projection, the aligned image slices can be transformed from a set of two dimensional images, $I_j(x,y)$, to a single three dimensional image, $I'(x,y,z)$. This three dimensional image is of particular interest when morphological information is required, further study can be undertaken using computer algorithms, such as isosurfaces and data slicing to analyse the data.

As TEM samples cannot typically be viewed at a full 180 degree rotation, the observed images typically suffer from a "missing wedge" of data. Mechanical techniques, such as multi-axis tilting, as well as numerical techniques exist to limit the impact of this missing data on the observed specimen morphology.

SAMPLE PREPARATION

The TEM is used heavily in material science, metallurgy and the biological sciences. In each case the specimens must be very thin and able to withstand the high vacuum present inside the instrument.

Sample preparation in TEM can be a complex procedure. TEM specimens are typically hundreds of nanometres thick, as the electron beam interacts readily with the sample, an effect that increases roughly with atomic number. High quality samples will have a thickness that is comparable to the mean free path of the electrons that travel through the samples, which may be only a few tens of nanometres. Preparation of TEM specimens is specific to the material under analysis and the desired information to obtain from the specimen. As such, many generic techniques have been used for the preparation of the required thin sections.

To withstand the instrument vacuum, biological specimens are typically held at liquid nitrogen temperatures after embedding in vitreous ice, or fixated using either a negative staining material such as uranyl acetate or by plastic embedding.

In material science and metallurgy the specimens tend to be naturally resistant to vacuum, but still must be prepared as a thin foil, or etched so some portion of the specimen is thin enough for the beam to penetrate. Constraints on the thickness of the material may be limited by the scattering cross-section of the atoms from which the material is comprised. Materials that have dimensions small enough to be electron transparent, such as powders or nanotubes, can be quickly produced by the deposition of a dilute sample containing the specimen onto support grids or films.

Tissue Sectioning

By passing samples over a glass or diamond edge, small, thin sections can be readily obtained using a semi-automated method This method is used to obtain thin, minimally deformed samples that allow for the observation of tissue samples. Additionally inorganic samples have been studied, such as aluminium, although this usage is limited owing to the heavy damage induced in the less soft samples.

Sample Staining

Details in light microscope samples can be enhanced by stains that absorb light. Similarly TEM samples, usually of biological tissues, can utilize stains to enhance contrast. The stain absorbs electrons or scatter out of the part of the electron beam that is projected onto the imaging system. Compounds of heavy metals such as osmium, lead, or uranium may be used prior to TEM observation to selectively deposit electron dense atoms in or on the sample in desired cellular or protein regions.

Mechanical Milling

Mechanical polishing may be used to prepare samples. Polishing needs to be done to a high quality, to ensure constant sample thickness across the region of interest. A diamond, or cubic boron nitride polishing compound may be used in the final stages of polishing to remove any scratches that may cause contrast fluctuations due to varying sample thickness. Even after careful mechanical milling, additional fine methods such as ion etching may be required to perform final stage thinning.

Chemical Etching

Certain samples may be prepared by chemical etching, particularly metallic specimens. These samples are thinned using a chemical etchant, such as an acid, to prepare the sample for TEM observation. Devices to control the thinning process may allow the operator to control either the voltage or current passing through the specimen, and may include systems to detect when the sample has been thinned to a sufficient level of optical transparency.

ION Etching

Ion etching is a sputtering process that can remove very fine quantities of material. This is used to perform a finishing polish of specimens polished by other means. Ion etching uses an inert gas passed through an electric field to generate a plasma stream that is directed to the sample surface. Acceleration energies for gases such as argon are typically a few kilovolts. The sample may be rotated to promote even polishing of the sample surface. The sputtering rate of such methods are on the order of tens of micrometres per hour, limiting the method to only extremely fine polishing.

More recently Focussed ion beam methods have been used to prepare samples. FIB is a relatively new technique to prepare thin samples for TEM examination from larger specimens. Because FIB can be used to micro-machine samples very precisely, it is possible to mill very thin membranes from a specific area of interest in a sample, such as a semiconductor or metal. Unlike intert gas ion sputtering, FIB makes use of significantly more energetic gallium ions and may alter the composition or structure of the material through gallium implantation.

Modifications

The capabilities of the TEM can be further extended by additional stages and detectors, sometimes incorporated on the same microscope. An *electron cryomicroscope* is a TEM with a specimen holder capable of maintaining the specimen at liquid nitrogen or liquid helium temperatures. This allows imaging specimens prepared in vitreous ice, the preferred preparation technique for imaging individual molecules or macromolecular assemblies.

A TEM can be modified into a scanning transmission electron microscope (STEM) by the addition of a system that rasters the beam across the sample to form the image, combined with suitable detectors.

An analytical TEM is one equipped with detectors that can determine the elemental composition of the specimen by analyzing its X-ray spectrum or the energy-loss spectrum of the transmitted electrons.

Modern research TEMs may include aberration correctors, to reduce the amount of distortion in the image, allowing information on features on the scale of 0.1 nm to be obtained (resolutions down to 0.05 nm have been achieved at magnifications of 50 million times Incident beam Monochromators may also be used which reduce the energy spread of the incident electron beam to less than 0.15 eV. Major TEM makers include JEOL, Hitachi High-technologies, FEI Company (from merging with Philips Electron Optics) and Carl Zeiss.

LIMITATIONS

There are a number of drawbacks to the TEM technique. Many materials require extensive sample preparation to produce a sample thin enough to be electron transparent, which makes TEM analysis a relatively time consuming process with a low throughput of samples. Graphene, a carbon nanomaterial, relatively transparent, very hard and just one atom thick, is currently being used as a platform on which the materials to be examined are placed. Being almost transparent to electrons, a graphene substrate has been able to show single hydrogen atom and hydrocarbons. The structure of the sample may also be changed during the preparation process. Also the field of view is relatively small, raising the possibility that the region analyzed may not be characteristic of the whole sample. There is potential that the sample may be damaged by the electron beam, particularly in the case of biological materials.

RESOLUTION LIMITS

Resolution of the HRTEM is limited by spherical and chromatic aberration, but a new generation of aberration correctors has been able to overcome spherical aberration Software correction of spherical aberration has allowed the production of images with sufficient resolution to show carbon atoms in diamond separated by only 0.89 Ångströms and atoms in silicon at 0.78 Ångströms (78 pm) at magnifications of 50 million times. Improved resolution has also allowed the imaging of lighter atoms that scatter electrons less efficiently—lithium atoms have been imaged in lithium battery materials The ability to determine the positions of atoms within materials has made the HRTEM an indispensable tool for nanotechnology research and development in many fields, including heterogeneous catalysis and the development of semiconductor devices for electronics and photonics.

CHAPTER - 16

Separation Process

In chemistry and chemical engineering, a separation process is used to transform a mixture of substances into two or more distinct products. The separated products could differ in chemical properties or some physical property, such as size, or crystal modification or other separation into different components.

Barring a few exceptions, almost every element or compound is found naturally in an impure state such as a mixture of two or more substances. Many times the need to separate it into its individual components arises. Separation applications in the field of chemical engineering are very important. A good example is that of crude oil. Crude oil is a mixture of various hydrocarbons and is valuable in this natural form. Demand is greater, however, for the purified various hydrocarbons such as natural gases, gasoline, diesel, jet fuel, lubricating oils, asphalt, etc.

Separation processes can essentially be termed as mass transfer processes. The classification can be based on the means of separation, *mechanical* or *chemical*. The choice of separation depends on the pros and cons of each. *Mechanical* separations are usually favored if possible due to the lower cost of the operations as compared to *chemical* separations.

Systems that can not be separated by purely mechanical means (e.g. crude oil), chemical separation is the remaining solution. The mixture at hand could exist as a combination of any two or more states: solid-solid, solid-liquid, solid-gas, liquid-liquid, liquid-gas, gas-gas, solid-liquid-gas mixture, etc.

Depending on the raw mixture, various processes can be employed to separate the mixtures. Many times two or more of these processes have to be used in combination to obtain the desired separation. In addition to *chemical* processes, *mechanical* processes can also be applied where possible. In the example of crude oil, one upstream distillation operation will feed its two or more product streams into multiple downstream distillation operations to further separate the crude, and so on until final products are purified.

CRYSTALLIZATION

Crystallization is the (natural or artificial) process of formation of solid crystals precipitating from a solution, melt or more rarely deposited directly from a gas. Crystallization is also a chemical solid-liquid separation technique, in which mass transfer of a solute from the liquid solution to a pure solid crystalline phase occurs.

Process

The crystallization process consists of two major events, *nucleation* and *crystal growth. Nucleation* is the step where the solute molecules dispersed in the solvent start to gather into clusters, on the nanometer scale (elevating solute concentration in a small region), that becomes stable under the current operating conditions. These stable clusters constitute the nuclei. However when the clusters are not stable, they redissolve. Therefore, the clusters need to reach a critical size in order to become stable nuclei. Such critical size is dictated by the operating conditions (temperature, supersaturation, etc.). It is at the stage of nucleation that the atoms arrange in a defined and periodic manner that defines the crystal structure — note that "crystal structure" is a special term that refers to the relative arrangement of the atoms, not the macroscopic properties of the crystal (size and shape), although those are a result of the internal crystal structure.

The *crystal growth* is the subsequent growth of the nuclei that succeed in achieving the critical cluster size. Nucleation and growth continue to occur simultaneously while the supersaturation exists. Supersaturation is the driving force of the crystallization, hence the rate of nucleation and growth is driven by the existing supersaturation in the solution. Depending upon the conditions, either nucleation or growth may be predominant over the other, and as a result, crystals with different sizes and shapes are obtained (control of crystal size and shape constitutes one of the main challenges in industrial manufacturing, such as for pharmaceuticals). Once the supersaturation is exhausted, the solid-liquid system reaches equilibrium and the crystallization is complete, unless the operating conditions are modified from equilibrium so as to supersaturate the solution again.

Many compounds have the ability to crystallize with different crystal structures, a phenomenon called polymorphism. Each polymorph is in fact a different thermodynamic solid state and crystal polymorphs of the same compound exhibit different physical properties, such as dissolution rate, shape (angles between facets and facet growth rates), melting point, etc. For this reason, polymorphism is of major importance in industrial manufacture of crystalline products.

Crystallization in nature

There are many examples of natural process that involve crystallization.

Geological time scale process examples include:

- Natural (mineral) crystal formation (see also gemstone);

- Stalactite/stalagmite, rings formation.

Usual time scale process examples include:

- Snow flakes formation (see also Koch snowflake);
- Honey crystallization (nearly all types of honey crystallize).

Artificial Methods

For crystallization (see also recrystallization) to occur from a solution it must be supersaturated. This means that the solution has to contain more solute entities (molecules or ions) dissolved than it would contain under the equilibrium (saturated solution). This can be achieved by various methods, with:

1. solution cooling;
2. addition of a second solvent to reduce the solubility of the solute (technique known as antisolvent or drown-out);
3. chemical reaction; and
4. change in pH being the most common methods used in industrial practice.

Other methods, such as solvent evaporation, can also be used.

Applications

There are two major groups of applications for the *artificial crystallization* process: *crystal production* and purification.

Crystal production

- Macroscopic crystal production, for supply the demand of natural-like crystals with methods that "accelerate time-scale" for massive production and/or perfection:
- ionic crystal production;

- covalent crystal production.
- Tiny size crystals.
- Powder, sand and smaller sizes: using methods for powder and controlled (nanotechnology fruits) forms.
- *Mass-production*: on chemical industry, like salt-powder production.
- *Sample production*: small production of tiny crystals for material characterization. Controlled recrystallization is an important method to supply unusual crystals, that are needed to reveal the molecular structure and nuclear forces inside a typical molecule of a crystal. Many techniques, like X-ray crystallography and NMR spectroscopy, are widely used in chemistry and biochemistry to determine the structures of an immense variety of molecules, including inorganic compounds and bio-macromolecules.
- Thin film production.

Massive production examples:

- "Powder salt for food" industry;
- Silicon crystal wafer production.
- Production of sucrose from sugar beet, where the sucrose is crystallized out from an aqueous solution.

Purification

Well formed crystals are expected to be pure because each molecule or ion must fit perfectly into the lattice as it leaves the solution. Impurities would normally not fit as well in the lattice, and thus remain in solution preferentially. Hence, molecular recognition is the principle of purification in crystallization. However, there are instances when impurities incorporate into the lattice, hence, decreasing the level of purity of the final crystal product. Also, in some cases, the solvent may incorporate into the lattice forming a *solvate*. In addition, the solvent may be 'trapped' (in liquid state) within the crystal formed, and this phenomenon is known as *inclusion*.

Thermodynamic view

The nature of a crystallization process is governed by both thermodynamic and kinetic factors, which can make it highly variable and difficult to control. Factors such as impurity level, mixing regime, vessel design, and cooling profile can have a major impact on the size, number, and shape of crystals produced.

Now put yourself in the place of a molecule within a pure and *perfect crystal*, being heated by an external source. At some sharply defined temperature, a bell rings, you must leave your neighbours, and the complicated architecture of the crystal collapses to that of a liquid. Textbook thermodynamics says that melting occurs because the entropy, S, gain in your system by spatial randomization of the molecules has overcome the enthalpy, H, loss due to breaking the crystal packing forces:

$$T(S_{liquid} - S_{solid}) > H_{liquid} - H_{solid}$$

$$G_{liquid} < G_{solid}$$

This rule suffers no exceptions when the temperature is rising. By the same token, on cooling the melt, at the very same temperature the bell should ring again, and molecules should click back into the very same crystalline form. The entropy decrease due to the ordering of molecules within the system is overcompensated by the thermal randomization of the surroundings, due to the release of the heat of fusion; the entropy of the universe increases.

But liquids that behave in this way on cooling are the exception rather than the rule; in spite of the second principle of thermodynamics, crystallization usually occurs at lower temperatures (supercooling). This can only mean that a crystal is more easily destroyed than it is formed. Similarly, it is usually much easier to dissolve a perfect crystal in a solvent than to grow again a good crystal from the resulting solution. The nucleation and growth of a crystal are under kinetic, rather than thermodynamic, control.

Equipment for Crystallization

1. *Tank crystallizers.* Tank crystallization is an old method still used in some specialized cases. Saturated solutions, in tank crystallization, are allowed to cool in open tanks. After a period of time the mother liquid is drained and the crystals removed. Nucleation and size of crystals are difficult to control. Typically, labor costs are very high.

2. *Scraped surface crystallizers.* One type of scraped surface crystallizer is the Swenson-Walker crystallizer, which consists of an open trough 0.6m wide with a semicircular bottom having a cooling jacket outside. A slow-speed spiral agitator rotates and suspends the growing crystals on turning. The blades pass close to the wall and break off any deposits of crystals on the cooled wall. The product generally has a somewhat wide crystal-size distribution.

3. *Double-pipe scraped surface crystallizer.* Also called a *votator,* this type of crystallizer is used in crystallizing ice cream and plasticizing margarine. Cooling water passes in the annular space. An internal agitator is fitted with spring-loaded scrapers that wipe the wall and provide good heat-transfer coefficients.

4. *Circulating-liquid evaporator-crystallizer.* Also called *Oslo crystallizer.* Here supersaturation is reached by evaporation. The circulating liquid is drawn by the screw pump down inside the tube side of the condensing stream heater. The heated liquid then flows into the vapor space, where flash evaporation occurs, giving some supersaturation.The vapor leaving is condensed. The supersaturated liquid flows down the

downflow tube and then up through the bed of fluidized and agitated crystals, which are growing in size. The leaving saturated liquid then goes back as a recycle stream to the heater, where it is joined by the entering fluid. The larger crystals settle out and slurry of crystals and mother liquid is withdrawn as a product.

5. *Circulating-magma vacuum crystallizer.* The magma or suspension of crystals is circulated out of the main body through a circulating pipe by a screw pump. The magma flows though a heater, where its temperature is raised 2-6 K. The heated liquor then mixes with body slurry and boiling occurs at the liquid surface. This causes supersaturation in the swirling liquid near the surface, which deposits in the swirling suspended crystals until they leave again via the circulating pipe. The vapors leave through the top. A steam-jet ejector provides vacuum.

6. *Continuous oscillatory baffled crystallizer (COBCTM).* The COBCTM is a tubular baffled crystallizer that offers plug flow under laminar flow conditions (low flow rates) with superior heat transfer coefficient, allowing controlled cooling profiles, e.g. linear, parobolic, discontinued, step-wise or any type, to be achieved. This gives much better control over crystal size, morphology and consistent crystal products. For further information see oscillatory baffled reactor.

DISTILLATION

Distillation is a method of separating mixtures based on differences in their volatilities in a boiling liquid mixture. Distillation is a unit operation, or a physical separation process, and not a chemical reaction.

Commercially, distillation has a number of uses. It is used to separate crude oil into more fractions for specific uses such as transport, power generation and heating. Water is distilled to remove impurities, such as salt from seawater. Air is distilled to separate its components—notably oxygen, nitrogen, and argon—for industrial use. Distillation of fermented solutions has been used since ancient times to produce distilled beverages with a higher alcohol content. The premises where distillation is carried out, especially distillation of alcohol, are known as a distillery.

History

Early types of distillation were known to the Babylonians in Mesopotamia (in what is now Iraq) from at least the 2nd millennium BC. Archaeological excavations in northwest Pakistan have yielded evidence that the distillation of alcohol was known in Pakistan since 500 BC but only became common between 150 BC - 350 AD. Distillation was later known to Greek alchemists from the 1st century AD and the later development of large-scale distillation apparatus occurred in response to demands for spirits According to K. B. Hoffmann the earliest mention of "destillatio per descensum" occurs in the writings of Aetius, a Greek physician from the 5th century Hypatia of Alexandria is credited with having invented an early distillation apparatus and the first clear description of early apparatus for distillation is given by Zosimos of Panopolis in the fourth century Primitive tribes of India used a method of distillation for producing *Mahuda* liquor. This crude and ancient method is not very effective.

The invention of highly effective "pure distillation" is credited to Arabic and Persian chemists in the Middle East from the 8th century. They produced distillation processes to isolate and purify chemical substances for industrial purposes such as isolating natural esters (perfumes) and producing pure alcohol. The first among them was Jabir ibn Hayyan (Geber), in the 8th century, who is credited with the invention of numerous chemical apparatus and processes that are still in use today. In particular, his alembic was the first still with retorts which could fully purify chemicals, a precursor to the pot still, and its design has served as inspiration for modern micro-scale distillation apparatus such as the Hickman stillhead. The isolation of ethanol (alcohol) as a pure compound through distillation was first achieved by the Arab chemist Al-Kindi (Alkindus). Petroleum was first distilled by the Persian alchemist Muhammad ibn Zakariya Razi (Rhazes) in the 9th century, for producing kerosene, while steam distillation was invented by Avicenna in the early 11th century, for producing essential oils.

As the works of Middle Eastern scribes made their way to India and became a part of Indian alchemy, several texts dedicated to distillation made their way to Indian libraries. Among these was a treatise written by a scholar from Bagdad in 1034 titled *Ainu-s-Sana'ah wa' Auna-s-Sana'ah*. Scholar Al-Jawbari travelled to India. By the time of the writing of the *Ain-e-Akbari*, the process of distillation was well known in India.

Distillation was introduced to medieval Europe through Latin translations of Arabic chemical treatises in the 12th century In 1500, German alchemist Hieronymus Braunschweig published *Liber de arte destillandi* (The Book of the Art of Distillation) the first book solely dedicated to the subject of distillation, followed in 1512 by a much expanded version. In 1651, John French published The Art of Distillation the first major English compendium of practice, though it has been claimed that much of it derives from Braunschweig's work. This includes diagrams with people in them showing the industrial rather than bench scale of the operation.

As alchemy evolved into the science of chemistry, vessels called retorts became used for distillations. Both alembics and retorts are forms of glassware with long necks pointing to the side at a downward angle which acted as air-cooled condensers to condense the distillate and let it drip downward for collection. Later, copper alembics were invented. Riveted joints were often kept tight by using various mixtures, for instance a dough made of rye flour These alembics often featured a cooling system around the beak, using cold water for instance, which made the condensation of alcohol more efficient. These were called pot stills. Today, the retorts and pot stills have been largely supplanted by more efficient distillation methods in most industrial processes. However, the pot still is still widely used for the elaboration of some fine alcohols such as cognac, Scotch whisky, tequila and some vodkas. Pot stills made of various materials (wood, clay, stainless steel) are also used by bootleggers in various countries. Small pot stills are also sold for the domestic production of flower water or essential oils.

Early forms of distillation were batch processes using one vaporization and one condensation. Purity was improved by further distillation of the condensate. Greater volumes were processed by simply repeating the distillation. Chemists were reported to carry out as many as 500 to 600 distillations in order to obtain a pure compound

In the early 19th century the basics of modern techniques including pre-heating and reflux were developed, particularly by the French , then in 1830 a British Patent was issued to Aeneas Coffey for a whiskey distillation column which worked continuously and may be regarded as the archetype of modern petrochemical units. In 1877, Ernest Solvay was granted a U.S. Patent for a tray column for ammonia distillation and the same and subsequent years saw developments of this theme for oil and spirits.

With the emergence of chemical engineering as a discipline at the end of the 19th century, scientific rather than empirical methods could be applied. The developing petroleum industry in the early 20th century provided the impetus for the development of accurate design methods such as the McCabe-Thiele method and the Fenske equation.

The availability of powerful computers has also allowed direct computer simulation of distillation columns.

Applications of Distillation

The application of distillation can roughly be divided in four groups: laboratory scale, industrial distillation, distillation of herbs for perfumery and medicinals (herbal distillate), and food processing. The latter two are distinct from the former two in that the distillation is not used as a true purification method but rather to transfer all volatiles from the source materials to the distillate.

The main difference between laboratory scale distillation and industrial distillation is that laboratory scale distillation is often performed batch-wise, whereas industrial distillation often occurs continuously. In batch distillation, the composition of the source material, the vapors of the distilling compounds and the distillate change during the distillation. In batch distillation, a still is charged (supplied) with a batch of feed mixture, which is then separated into its component fractions which are collected sequentially from most volatile to less volatile, with the bottoms (remaining least or non-volatile fraction) removed at the end. The still can then be recharged and the process repeated.

In continuous distillation, the source materials, vapors, and distillate are kept at a constant composition by carefully replenishing the source material and removing fractions from both vapor and liquid in the system. This results in a better control of the separation process.

Idealized Distillation Model

The boiling point of a liquid is the temperature at which the vapor pressure of the liquid equals the pressure in the liquid, enabling bubbles to form without being crushed. A special case is the normal boiling point, where the vapor pressure of the liquid equals the ambient atmospheric pressure.

It is a common misconception that in a liquid mixture at a given pressure, each component boils at the boiling point corresponding to the given pressure and the vapors of each component will collect separately and purely. This, however, does not occur even in an idealized system. Idealized models of distillation are essentially governed by Raoult's law and Dalton's law, and assume that vapor-liquid equilibria are attained.

Raoult's law assumes that a component contributes to the total vapor pressure of the mixture in proportion to its percentage of the mixture and its vapor pressure when pure, or succinctly: partial pressure equals mole fraction multiplied by vapor pressure when pure. If one component changes another component's vapor pressure, or if the volatility of a component is dependent on its percentage in the mixture, the law will fail.

Dalton's law states that the total vapor pressure is the sum of the vapor pressures of each individual component in the mixture. When a multi-component liquid is heated, the vapor pressure of each component will rise, thus causing the total vapor pressure to rise. When the total vapor pressure reaches the pressure surrounding the liquid, boiling occurs and liquid turns to gas throughout the bulk of the liquid. Note that a mixture with a given composition has one boiling point at a given pressure, when the components are mutually soluble.

An implication of one boiling point is that lighter components never cleanly "boil first". At boiling point, all volatile components boil, but for a component, its percentage in the vapor is the same as its percentage of the total vapor pressure. Lighter components have a higher partial pressure and thus are concentrated in the vapour, but heavier volatile components also have a (smaller) partial pressure and necessarily evaporate also, albeit being less concentrated in the vapor. Indeed, batch distillation and fractionation succeed by varying the composition of the mixture. In batch distillation, the batch evaporates, which changes its composition; in fractionation, liquid higher in the fractionation column contains more lights and boils at lower temperatures.

The idealized model is accurate in the case of chemically similar liquids, such as benzene and toluene. In other cases, severe deviations from Raoult's law and Dalton's law are observed, most famously in the mixture of ethanol and water. These compounds, when heated together, form an azeotrope, which is a composition with a boiling point higher or lower than the boiling point of each separate liquid. Virtually all liquids, when mixed and heated, will display azeotropic behaviour. Although there are computational methods that can be used to estimate the behavior of a mixture of arbitrary components, the only way to obtain accurate vapor-liquid equilibrium data is by measurement.

It is not possible to *completely* purify a mixture of components by distillation, as this would require each component in the mixture to have a zero partial pressure. If ultra-pure products are the goal, then further chemical separation must be applied. When a binary mixture is evaporated and the other component, e.g. a salt, has zero partial pressure for practical purposes, the process is simpler and is called evaporation in engineering.

Heating an ideal mixture of two volatile substances A and B (with A having the higher volatility, or lower boiling point) in a batch distillation setup (such as in an apparatus depicted in the opening figure) until the mixture is boiling results in a vapor above the liquid which contains a mixture of A and B. The ratio between A and B in the vapor will be different from the ratio in the liquid: the ratio in the liquid will be determined by how the original mixture was prepared, while the ratio in the vapour will be enriched in the more volatile compound, A (due to Raoult's Law, see above). The vapor goes through the condenser and is removed from the system. This in turn means that the ratio of compounds in the remaining liquid is now different from the initial ratio (i.e. more enriched in B than the starting liquid).

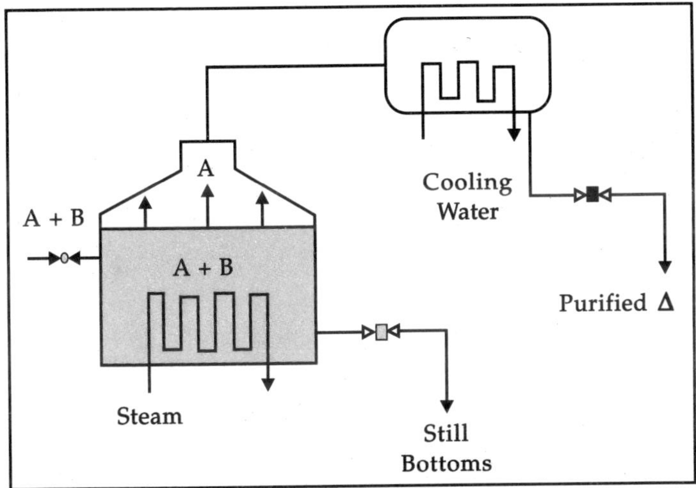

Fig 16.1: A batch still showing the separation of A and B

BATCH DISTILLATION

The result is that the ratio in the liquid mixture is changing, becoming richer in component B. This causes the boiling point of the mixture to rise, which in turn results in a rise in the temperature in the vapour, which results in a changing ratio of A: B in the gas phase (as distillation continues, there is an increasing proportion of B in the gas phase). This results in a slowly changing ratio A: B in the distillate.

If the difference in vapour pressure between the two components A and B is large (generally expressed as the difference in boiling points), the mixture in the beginning of the distillation is highly enriched in component A, and when component A has distilled off, the boiling liquid is enriched in component B.

Continuous Distillation

Continuous distillation is an ongoing distillation in which a liquid mixture is continuously (without interruption) fed into the process and separated fractions are removed continuously as output streams as time passes during the operation. Continuous distillation produces at least two output fractions, including at least one volatile distillate fraction, which has boiled and been separately captured as a vapor condensed to a liquid. There is always a bottoms (or residue) fraction, which is the least volatile residue that has not been separately captured as a condensed vapor.

Continuous distillation differs from batch distillation in the respect that concentrations should not change over time. Continuous distillation can be run at a steady state for an arbitrary amount of time. Given a feed of in a specified composition, the main variables that affect the purity of products in continuous distillation are the reflux ratio and the number of theoretical equilibrium stages (practically, the number of trays or the height of packing). Reflux is a flow from the condenser back to the column, which generates a recycle that allows a better separation with a given number of trays. Equilibrium stages are ideal steps where compositions achieve vapor-liquid equilibrium, repeating the separation process and allowing better separation given a reflux ratio. A column with a high reflux ratio may have fewer stages, but it refluxes a large amount of liquid, giving a wide column with a large holdup. Conversely, a column with a low reflux ratio must have a large number of stages, thus requiring a taller column.

Continuous distillation requires building and configuring dedicated equipment. The resulting high investment cost restricts its use to the large scale.

General Improvements

Both batch and continuous distillations can be improved by making use of a fractionating column on top of the distillation flask. The column improves separation by providing a larger surface area for the vapor and condensate to come into contact. This helps it remain at equilibrium for as long as possible. The column can even consist of small subsystems ('trays' or 'dishes') which all contain an enriched, boiling liquid mixture, all with their own vapor-liquid equilibrium.

There are differences between laboratory-scale and industrial-scale fractionating columns, but the principles are the same. Examples of laboratory-scale fractionating columns (in increasing efficacy) include:

- Packed column (packed with glass beads, metal pieces, or other chemically inert material);
- Spinning band distillation system.

Laboratory Scale Distillation

Laboratory scale distillations are almost exclusively run as batch distillations. The device used in distillation, sometimes referred to as a *still*, consists at a minimum of a reboiler or *pot* in which the source material is heated, a condenser in which the heated vapour is cooled back to the liquid state, and a receiver in which the concentrated or purified liquid, called the distillate, is collected. Several laboratory scale techniques for distillation exist (see also distillation types).

Simple Distillation

In simple distillation, all the hot vapors produced are immediately channeled into a condenser which cools and condenses the vapors. Therefore, the distillate will not be pure - its composition will be identical to the composition of the vapors at the given temperature and pressure, and can be computed from Raoult's law.

As a result, simple distillation is usually used only to separate liquids whose boiling points differ greatly (rule of thumb is 25 °C), or to separate liquids from involatile solids or oils. For these cases, the vapor pressures of the components are usually sufficiently different that Raoult's law may be neglected due to the insignificant contribution of the less volatile component. In this case, the distillate may be sufficiently pure for its intended purpose.

Fractional Distillation

For many cases, the boiling points of the components in the mixture will be sufficiently close that Raoult's law must be taken into consideration. Therefore, fractional distillation must be used in order to separate the components well by repeated vaporization-condensation cycles within a packed fractionating column. This separation, by successive distillations, is also referred to as rectification.

As the solution to be purified is heated, its vapors rise to the fractionating column. As it rises, it cools, condensing on the condenser walls and the surfaces of the packing material. Here, the condensate continues to be heated by the rising hot vapors; it vaporizes once more. However, the composition of the fresh vapors are determined once again by Raoult's law. Each vaporization-condensation cycle (will yield a purer solution of the more volatile component. In reality, each cycle at a given temperature does not occur at exactly the same position in the fractionating column; *theoretical plate* is thus a concept rather than an accurate description.

More theoretical plates lead to better separations. A spinning band distillation system uses a spinning band of Teflon or metal to force the rising vapors into close contact with the descending condensate, increasing the number of theoretical plates.

Steam Distillation

Like vacuum distillation, steam distillation is a method for distilling compounds which are heat-sensitive. This process involves using bubbling steam through a heated mixture of the raw material. By Raoult's law, some of the target compound will vaporize (in accordance with its partial pressure). The vapor mixture is cooled and condensed, usually yielding a layer of oil and a layer of water.

Steam distillation of various aromatic herbs and flowers can result in two products; an essential oil as well as a watery herbal distillate. The essential oils are often used in perfumery and aromatherapy while the watery distillates have many applications in aromatherapy, food processing and skin care.

Vacuum Distillation

Some compounds have very high boiling points. To boil such compounds, it is often better to lower the pressure at which such compounds are boiled instead of increasing the temperature. Once the pressure is lowered to the vapor pressure of the compound (at the given temperature), boiling and the rest of the distillation process can commence. This technique is referred to as vacuum distillation and it is commonly found in the laboratory in the form of the rotary evaporator.

This technique is also very useful for compounds which boil beyond their decomposition temperature at atmospheric pressure and which would therefore be decomposed by any attempt to boil them under atmospheric pressure.

Molecular distillation is vacuum distillation below the pressure of 0.01 torr In fact, 0.01 torr is rarefied medium vacuum or only one order of magnitude more rarefied than high vacuum, where the mean free path of molecules is comparable to the size of the equipment. The gaseous phase no longer exerts significant pressure on the substance to be evaporated, and consequently, rate of evaporation no longer depends on pressure. That is, because the continuum assumptions of fluid dynamics no longer apply, mass transport is governed by molecular dynamics rather than fluid dynamics. Thus, a short path between the hot surface and the cold surface is necessary. Molecular distillation is used industrially for purification of oils.

Air-sensitive Vacuum Distillation

Some compounds have high boiling points as well as being air sensitive. A simple vacuum distillation system as exemplified above can be used, whereby the vacuum is replaced with an inert gas after the distillation is complete. However, this is a less satisfactory system if one desires to collect fractions under a reduced pressure. To do this a "pig" adaptor can be added to the end of the condenser, or for better results or for very air sensitive compounds a Perkin triangle apparatus can be used.

The Perkin triangle, has means via a series of glass or Teflon taps to allows fractions to be isolated from the rest of the still, without the main body of the distillation being removed from either the vacuum or heat source, and thus can remain in a state of reflux. To do this, the sample is first isolated from the vacuum by means of the taps, the vacuum over the sample is then replaced with an inert gas (such as nitrogen or argon) and can then be stoppered and removed. A fresh collection vessel can then be added to the system, evacuated and linked back into the distillation system via the taps to collect a second fraction, and so on, until all fractions have been collected.

Short Path Distillation

Short path distillation is a distillation technique that involves the distillate traveling a short distance, often only a few centimeters. A classic example would be a distillation involving the distillate traveling from one glass bulb to another, without the need for a condenser separating the two chambers. This technique is often used for compounds which are unstable at high temperatures. The advantage is that the heating temperature can be considerably lower (at this reduced pressure) than the boiling point of the liquid at standard pressure, and that the distillate only has to travel a short distance before condensing. The Kugelrohr is a kind of a short path distillation apparatus.

Other Types

- The process of *reactive distillation* involves using the reaction vessel as the still. In this process, the product is usually significantly lower-boiling than its reactants. As the product is formed from the reactants, it is vaporized and removed from the reaction mixture. This technique is an example of a continuous vs. a batch process; advantages include less downtime to charge the reaction vessel with starting material, and less workup.
- *Pervaporation* is a method for the separation of mixtures of liquids by partial vaporization through a non-porous membrane.
- *Extractive distillation* is defined as distillation in the presence of a miscible, high boiling, relatively non-volatile component, the solvent, that forms no azeotrope with the other components in the mixture.
- *Flash evaporation* (or partial evaporation) is the partial vaporization that occurs when a saturated liquid stream undergoes a reduction in pressure by passing through a throttling valve or other throttling device. This process is one of the simplest unit operations, being equivalent to a distillation with only one equilibrium stage.
- *Codistillation* is distillation which is performed on mixtures in which the two compounds are not miscible.

The unit process of evaporation may also be called "distillation":

- In *rotary evaporation* a vacuum distillation apparatus is used to remove bulk solvents from a sample. Typically the vacuum is generated by a water aspirator or a membrane pump.
- In a kugelrohr a short path distillation apparatus is typically used (generally in combination with a (high) vacuum) to distill high boiling (> 300 °C) compounds.

The apparatus consists of an oven in which the compound to be distilled is placed, a receiving portion which is outside of the oven, and a means of rotating the sample. The vacuum is normally generated by using a high vacuum pump.

Other Uses

- *Dry distillation* or *destructive distillation*, despite the name, is not truly distillation, but rather a chemical reaction known as pyrolysis in which solid substances are heated in an inert or reducing atmosphere and any volatile fractions, containing high-boiling liquids and products of pyrolysis, are collected. The destructive distillation of wood to give methanol is the root of its common name - *wood alcohol*.
- *Freeze distillation* is an analogous method of purification using freezing instead of evaporation. It is not truly distillation, but a recrystallization where the product is the mother liquor, and does not produce products equivalent to distillation. This process is used in the production of ice beer and ice wine to increase ethanol and sugar content, respectively. Unlike distillation, freeze distillation of ferment concentrates poisonous congeners rather than removing them like distillation.

Azeotropic Distillation

Interactions between the components of the solution create properties unique to the solution, as most processes entail nonideal mixtures, where Raoult's law does not hold. Such interactions can result in a constant-boiling azeotrope which behaves as if it were a pure compound (i.e., boils at a single temperature instead of a range). At an azeotrope, the solution contains the given component in the same proportion as the vapour, so that evaporation does not change the purity, and distillation does not effect separation. For example, ethyl alcohol and water form an azeotrope of 95.6% at 78.1 °C.

If the azeotrope is not considered sufficiently pure for use, there exist some techniques to break the azeotrope to give a pure distillate. This set of techniques are known as azeotropic distillation. Some techniques achieve this by "jumping" over the azeotropic composition (by adding an additional component to create a new azeotrope, or by varying the pressure). Others work by chemically or physically removing or sequestering the impurity. For example, to purify ethanol beyond 95%, a drying agent or a (desiccant such as potassium carbonate) can be added to convert the soluble water into insoluble water of crystallization. Molecular sieves are often used for this purpose as well.

Immiscible liquids, such as water and toluene, easily form azeotropes. Commonly, these azeotropes are referred to as a low boiling azeotrope because the boiling point of the azeotrope is lower than the boiling point of either pure component. The temperature and composition of the azeotrope is easily predicted from the vapor pressure of the pure components, without use of Raoult's law. The azeotrope is easily broken in a distillation set-up by using a liquid-liquid separator (a decanter) to separate the two liquid layers that are condensed overhead. Only one of the two liquid layers is refluxed to the distillation set-up.

High boiling azeotropes, such as a 20 weight per cent mixture of hydrochloric acid in water, also exist. As implied by the name, the boiling point of the azeotrope is greater than the boiling point of either pure component.

To break azeotropic distillations and cross distillation boundaries, such as in the DeRosier Problem, it is necessary to increase the composition of the light key in the distillate.

Breaking an Azeotrope with Unidirectional Pressure Manipulation

The boiling points of components in an azeotrope overlap to form a band. By exposing an azeotrope to a vacuum or positive pressure, it's possible to bias the boiling point of one component away from the other by exploiting the differing vapour pressure curves of each; the curves may overlap at the azeotropic point, but are unlikely to be remain identical further along the pressure axis either side of the azeotropic point. When the bias is great enough, the two boiling points no longer overlap and so the azeotropic band disappears.

This method can remove the need to add other chemicals to a distillation, but it has two potential drawbacks.

Under negative pressure, power for a vacuum source is needed and the reduced boiling points of the distillates requires that the condenser be run cooler to prevent distillate vapours being lost to the vacuum source. Increased cooling demands will often require additional energy and possibly new equipment or a change of coolant.

Alternatively, if positive pressures are required, standard glassware can not be used, energy must be used for pressurization and there is a higher chance of side reactions occurring in the distillation, such as decomposition, due to the higher temperatures required to effect boiling.

A unidirectional distillation will rely on a pressure change in one direction, either positive or negative.

Pressure-swing Distillation

Pressure-swing distillation is essentially the same as the unidirectional distillation used to break azeotropic mixtures, but here both positive and negative pressures may be employed.

This has an important impact on the selectivity of the distillation and allows a chemist to optimize a process such that fewer extremes of pressure and temperature are required and less energy is consumed. This is particularly important in commercial applications.

Pressure-swing distillation is employed during the industrial purification of ethyl acetate after its catalytic synthesis from ethanol.

Index

D

E

F

G

H

I

N

O

P

Q

❑❑❑